Ethnobotany of Nasik District, *MAHARASHTRA*

Ethnobotany of Nasik District, *MAHARASHTRA*

M.V. Patil
M.Sc., Ph.D.
Lecturer in Botany,
G.E.T.'s Arts, Commerce and Science College,
Nagaon – 424004, District Dhule (M.S.)

D.A. Patil
M.Sc., Ph.D.
Reader in Botany,
Post-Graduate Department of Botany,
S.S.V.P.Sanstha's L.K.Dr.P.R.Ghogrey, Science College,
Dhule-424005 (M.S.)

2006
DAYA PUBLISHING HOUSE
Delhi - 110 035

ISBN 81-7035-438-2

Published by : **Daya Publishing House**
1123/74, Deva Ram Park
Tri Nagar, Delhi - 110 035
Phone: 27383999
Fax: (011) 23244987
e-mail : dayabooks@vsnl.com
website : www.dayabooks.com

Showroom : 4760-61/23, Ansari Road, Darya Ganj,
New Delhi - 110 002
Phone: 23245578, 23244987

Laser Typesetting : **Classic Computer Services**
Delhi - 110 035

Printed at : **Chawla Offset Printers**
Delhi - 110 052

PRINTED IN INDIA

Dedicated to the Memory of

Dr. V.D. Pathak

for his great vision in initiating Ethnobotanical Investigations in the State of Maharasthra

Preface

Generally, a layman raises three queries. What is the plant? Where it is found? And what is its use. It is the only science of ethnobotany that includes answers of all his questions together. We are also aware that plants constitute the life support system and make up irreplaceable resource base which ranges from aesthetics to food to medicine. Therefore, they have engaged the inquisitive attention of mankind from the time immemorable. Moreover, in modern times, the pressing need for natural products, the dire necessity to conserve genes, plants and ecosystems, and the realization of the importance of plants in conservation and improvement of environment, have led to the strengthening of our studies on indigenous plants and traditional knowledge. The threats to biodiversity and forces of acculturation are also warning to gear up studies on man-plant relationships. The present authors, therefore, took up the task of botanization in the district to attempt at a layman's queries in the Nasik District of Maharashtra. Previously, Nasik district has attracted attention of many botanists. They mostly paid attention to the identification of plant species in the region. Few stray attempts have been made to study the flora of the district from utilization point of view. We decided to study it selecting ethnobotany of the district as an exclusive topic of research, the results of which form the subject-matter of this book.

We are delighted in presenting our research work as we endeavoured to play role in developing an understanding and respect for man's alliances with plants in the economy of nature, so that they may be fostered. There must be an increased awareness of the facts of man's relationship with nature–the facts as they were, as they are, and as they could be in the future. This is the reason to publish the results of our studies, which in turn, will help in bringing about an awareness of these facts.

Forest management in India laid more emphasis on timber and marginal stress on non-wood forest products. Ethnobotanical studies carried out especially in last five decades in our country indicated how the forest-communities draw major segment of forest produce and are sustained on what is generally called 'minor' forest products. Ethnobotanical studies thus point out the factual situation helpful for forest management. Our study will help guide for appropriate extraction and utilization of MFP in Nasik District.

We tender our sincere thanks to the staff of Department of Forest, Nasik District, for their co-operation during our visits at different forest sites. We are also thankful to a large number tribal friends of the district for their help in passing their hereditary knowledge to us without hesitation and their ever time hospitality. It is our pleasant duty to record our deep sense of gratitude to Ex-Principal Shri B.M. Patil, S.S.V.P.S's L.K. Dr. P.R. Ghogrey Science College, Dhule for library and laboratory facilities. We are grateful to Principal Dr. P.N. Patil and Ex-Principal Shri S.S. Patil and I/C Principal Shri R.C. Chautmal for encouragements. We are also grateful to the librarians of Dr. B.A. Marathwada University, Aurangabad; Agharkar Research Institute (MACS), Pune, Botanical Survey of India, Western Circle, Pune; and Indian Agricultural Research Institute Library, New Delhi and for their help in library facilities.

We are indebted to Prof. Dr. R.M. Pai, Ex-Head, P.G. Department of Botany, Dr. B.A. Marathwada University, Aurangabad and Prof. Dr. M.H. Parabia, P.G. Department of Bio-Sciences, South Gujarat University, Surat, for their constant inspirations throughout the period of research.

Dr. D.A.Patil

Dr. M.V.Patil

Contents

Chapter 1
Introduction

The first interest of mankind in plant life dates back to the earliest times when, in view of his survival, he soon recognized the need to become familiar with the plants of his environment and engaged in food crop development. Concomitantly with the earliest need of plants for food, there was a keen awareness of the values of many forms of plant life as sources of medicine. This knowledge descended from generation to generation in the form of folklores.

Plant wealth is renewable resource and has substantially contributed to our economic development. It provides goods and services to the people and industry. Plants are interwoven with our culture. The vital role that plants play in the life and economy of all forest dwelling tribes, rural, semi-rural and urban population is well-known. They are traditionally providing food, medicare and other forest products for our daily needs. This knowledge has now emerged as a science of 'Ethnobotany'.

The term 'Ethnobotany' was first used by Harshberger (1885). Its scope was much elaborated later. This science of 'Ethnobotany' now in the present wider context denotes the entire realm of useful traditional relationship between plants and man. Ethnobotanical investigations have assumed great importance in enhancing our knowledge about the plants flourishing and used by native or aboriginal societies.

The Indian sub-continent is unique in the richness of its plant wealth. Over 15000 species of higher plants occur in our country. This floristic diversity is spread in different natural habitats. It is usually found in such areas that native communities dominate, use native plant wealth and practice traditional agriculture, in keeping with their own life styles, customs, rituals and beliefs. Our country is a land of amalgam of religions, cultures, races, languages and dialects, apart from its biological diversity. Interestingly, we have 563 kinds of aboriginals of diverse origin. This unparallel feature of our land, no wonder, attracts many of our research workers and scholars.

Western Ghats (Sahyadris) is a rich repository of vegetational wealth, the offshoots of which extend in Nasik District. The aboriginal communities like Mahadeo Koli, Kokani (Kunbi Kokna), Warli, Katkari, Bhil, Thakur etc. inhabit these difficult terrains. They have also their own customs, rituals and ceremonies. Their plantlore is also rich. Tribal and other ethnic communities continue to depend on plants for food, clothing, shelter and above all for alleviating diseases. They have elaborated their own traditional herbal cures, treasured in the native societies.

A perusal of literature, to date, warranted that this region is not fully documented ethnobotanically. It is either underexplored or certain pockets of it are practically unexplored. The earlier attempts particularly emphasized the floristic investigations. Ethnobotanical information, whatever is revealed, are largely a part of routine floristic studies in the region.

It was, therefore, felt worthwhile and at the suggestion of senior author Dr.D.A.Patil, Reader, Post-Graduate Department of Botany, S.S.V.P.Sanstha's L.K.Dr.P.R.Ghogrey Science College, Dhule (M.S.) to undertake fairly extensive and intensive ethnobotanical investigation as an exclusive topic of research in this district of Maharashtra.

Geography

Nasik District is located between 19°3′ and 20°52′ north latitude and 73°16′ and 74°56′ east longitude with an area of 15,582 sq.km

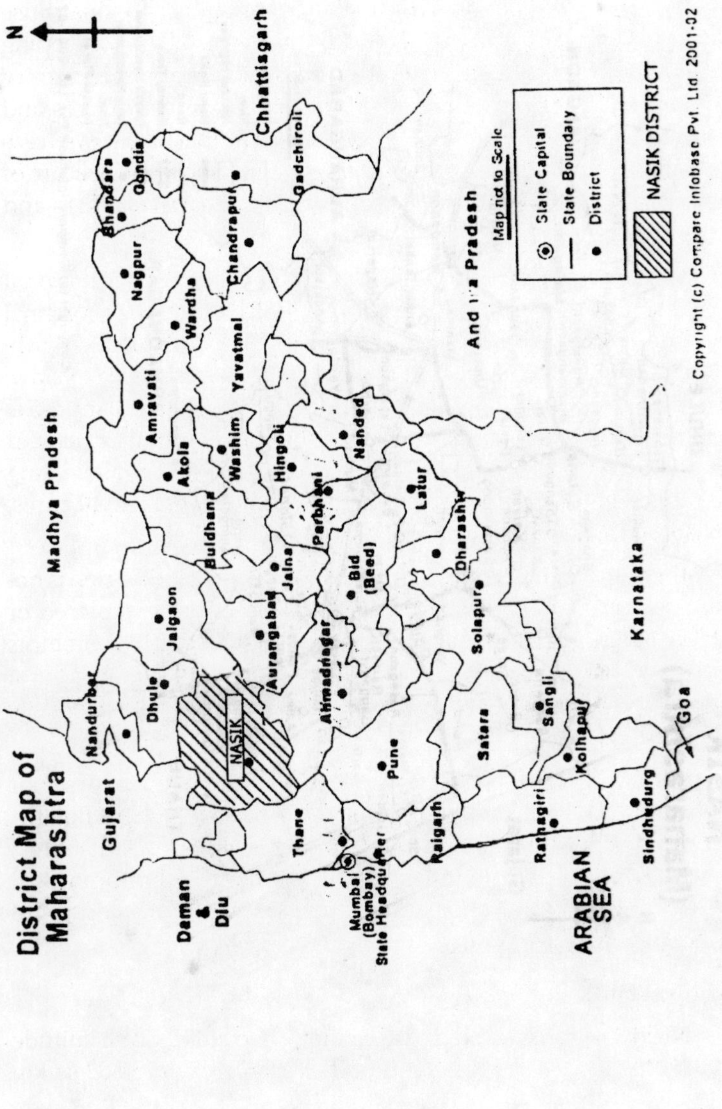

District Map of Maharashtra
N
Madhya Pradesh
Chhattisgarh
Gujarat
Daman & Diu
Andhra Pradesh
Karnataka
ARABIAN SEA
Goa
Nandurbar
Dhule
Jalgaon
NASIK
Aurangabad
Jalna
Buldhana
Akola
Washim
Amravati
Wardha
Nagpur
Bhandara
Gondia
Yavatmal
Chandrapur
Gadchiroli
Hingoli
Nanded
Parbhani
Latur
Bid (Beed)
Ahmadnagar
Thane
Mumbai (Bombay) State Headquarter
Pune
Raigarh
Satara
Solapur
Dharashiv
Sangli
Kolhapur
Ratnagiri
Sindhudurg
Map not to Scale
State Capital
State Boundaly
District
NASIK DISTRICT
Copyright (c) Compare Infobase Pvt. Ltd. 2001-02

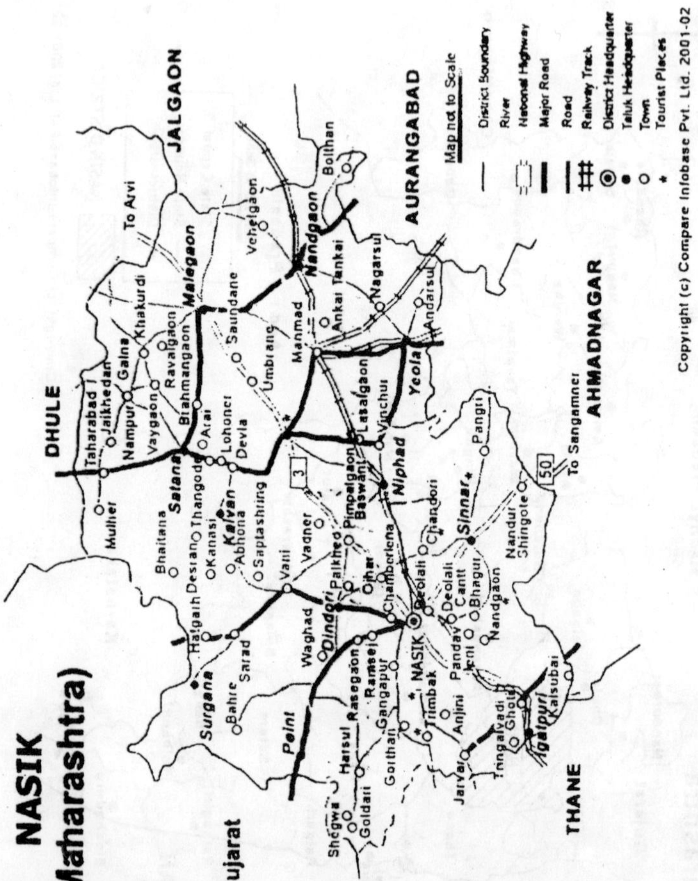
NASIK
(Maharashtra)
N
Gujarat
DHULE
JALGAON
AURANGABAD
AHMADNAGAR
THANE
To Arvi
To Sangamner
Map not to Scale
District Boundary
River
National Highway
Major Road
Road
Railway Track
District Headquarter
Taluk Headquarter
Town
Tourist Places
Copyright (c) Compare Infobase Pvt. Ltd. 2001-02

has population of 49,87,923. Nasik District is bounded on the north-west by Dangs and Surat Districts of Gujarat, on the north by Dhule district, on the east by Jalgaon and Aurangabad districts, on the south by Ahmednagar district and towards the south-west by Thane Districts (Map).

The district is divided into thirteen revenue talukas *viz.*, Satana (Baglan) Malegaon, Kalwan, Surgana, Peth, Dindori, Chandwad, Nandgaon, Niphad, Yeola, Nasik, Sinnar, Igatpuri (Map). The forests of district are divided into fifteen ranges grouped under two divisions *viz.*, East and West Nasik forest Divisions. The forests of district cover an area of 3,446.28 sq.km, of which 3,338.85 km are in charge of forest department, whereas 107.43 sq.km are governed by revenue department of Maharashtra state. The area under the control of former consist of reserved forest covering 2,920.08 sq.km. The protected forests covering 245.45 sq.km and the unclassified forests extending to 173.32 sq.km. The district has ca 20 towns and 1652 villages. The population of the district (2001 Census) is 49,87,923 out of which 9.31 lakhs are tribal people (*cf.* Anonymous, 2001a-b).

With the exception of Western Ghats, the general direction of mountain ranges is from west to east or south-west to north-east, the higher portions being near the west the ranges such as (1) Selbari range, (2) The Satmala range and (3) The Trimbak-Anjaneri range. These extend throughout the district. In the extreme north is the Selbari range. The higher points in it vary from about 1300 m in the west to half that height in the east. The highest peaks upto 1331 m are to be found in the Mangi-Tungi hills. The Satmala, Chandwad or Ajanta range runs right across the district. It differs from the rest of the mountains in north by number and shape of its peaks, and by the narrowing down of flat, summits. The highest peak in this range in 'Dhodap' (1451 m). Several other peaks approach this height, among these are Saptashring (1420 m), Indrai (1410 m) and Chandwad (1217 m). This range branches off from the Western Ghats (Sahyadris) in a easterly and south-easterly direction. With few spurs to the south, it throws off several branch spurs on nearly to the north-east as residual hills. The Trimbak-Anjaneri range stretches nearly eastwards from Bhaskerghad (1086 m) on the Western Ghats. Nearly 5 km east of it is Harish fort (1113 m) and Brahma Dongar (1201 m). The greatest heights are attained at Trimbak (1294 m) and Anjaneri (1300 m). This range includes the highest summits, one of which is Kalsubai (1846 m). The eastern and southern parts of the district are plain with general ground level being 500–550 m.

Geology and Soils

The great trap region covers the whole district. It is entirely of volcanic formation. The volcanic portion consists of compact stratified basalts and as early trap. The basalts are the most conspicuous geological feature. To the west, they lie in flat-topped ranges, separated by valleys, extending from west to east. In some flow the basalt is columnar and then weathers into fantastic shapes. The formation at the base of the traps in chiefly amygdaloidal, containing quarts in vertical veins, crystals and zealitic minerals, especially apophyllite weathering into a grey soil. The absence of laterite, which caps the summits of hills to the south, is a curious feature in the geology of area.

In between the flows, at places thin beds of volcanic tuffs are found. The flows are intruded by dykes. Along the courses of major rivers alluvium is found.

Deccan Traps

The deccan traps are upper cretaceous-lower eocene in age and are the products of fissure types of volcanic eruptions. The basaltic flow are two types *viz.*, pahoehoe and aa. The pahoehoe flows weather easily and gives rise to mature type of topography, with smooth hill slopes and conical peaks. Kalsubai hill exposes a 620.5 m thick pile of compound pahoehoe flows and one aa flow.

Tuffs

Thin beds of volcanic tuffs are found at places between the flows. Good exposure of tuffs are seen near mile stone 177 on Bombay-Agra highway.

Dykes

The basalts are intruded by a number of dykes in the Satana area.

Alluvium

Along the banks of Godavari, Girna and other rivers a few meter thick alluvium is deposited at places. This comprises of pebbles, sand and slit derived mostly from traps.

The valleys are filled with disintegrated basalt of various shades from grey to black. It is argillaceous in nature. The black soil contains high allumina and carbonates of calcium and magnesium with variable amount of nitrogen, potassium and phosphorus. The red soil is less common lime nodules and kanker are found practically. all over the district. Minor disseminations of native copper in the form of thin scales and ribbons are recorded in the vasicular cavity of a pahoehoe flow. Well developed and transparent crystals of quartz are seen in geodes and Vugs Kanker is found in alluvium and soil.

The River System

The district is drained by two chief rivers the Girna and the Godavari and their tributaries. The water shed between these are the Satmala range. Apart from this, there are number of small rivers draining westwards Konkan side into the Arabian sea. These are Chondi, Kavera, Sasu, Man, Nar, Par, Barik, Damanganga, Val, Vaitarna, Bhima, etc. The Girna rises just south of Cherai village at about 8 km south-west of Haatgad in the Western Ghats. It flows nearly east along a wide bed. It received in it to upper coarse several rivers of nearly equal size as itself. These are Tambadi, Punand, Aran, Mosum, Panjan, Maniyad and Bori. The Godavari or also called locally the Ganga, is the most celebrated river of the peninsular India. The principal course lies just below the side of the Trimbak. After passing the town of Trimbak, the Godavari turns to the west, cutting a deep and rocky bed through the Ghatmatha country. It receives the tributaries such as Kashyap, Nasardi, Darna, Kadva, Vaki, Unduhol, Valdevi, Dev, Jham, Banganga, Unanda and Gui.

Climate

The climatic conditions prevailing in the entire district is characterised by dryness except in the south-west monsoon seasons. The entire year may be divided into four seasons. The monsoon season starts from June and continues till the month of September. It is followed by post-monsoon season from October to November. The cold season starts from December and continues till the month of February. It is then followed by hot season which starts from March and ends in the month of May.

Rainfall

The south-west monsoon breaks by the middle of June, however, regular rainy season starts by the end of June and continues till the middle of October. In November and December occasional showers are also received in this region due to north-eastern monsoon (Table 1) gives the rainfall of last five years in the district.

Table 1: Rainfall (mm) in Nasik District (1996 to 2001)

Year	*Average Rainfall*
1996–1997	14643.2
1997–1998	16616.1
1998–1999	13241.8
1999–2000	8418.35
2000–2001	09619.04

On an average there are 51 rainy days in a year in the district. The average annual rainfall in the district is ca 1034.5 mm. The number of rainy days is high on the narrow strip of the district in close proximity to the Western Ghats. It varies from 89 days at Peth in the north to 102 days at Igatpuri in the south. Over the plateau region the number of rainy days decreases in general from the west to the east and varies from 32 at Satana to 54 at Dindori. About 88 per cent of the annual rainfall in the district is received during the south-west monsoon season. The highest precipitation is received in the month of July. During May and post-monsoon months some rainfall occur in the form of thunder showers.

Temperature

The western hilly part of the district enjoys a more equitable climate, while the eastern part of the district experiences extremes of climate with temperature lowering down to 5°C in winter and shooting up to 43°C, in summer. After February temperature in the district increases steadily till May. May is the hottest month with the mean daily maximum temperature being 40.6°C at Malegaon and 37.4°C at Nasik. The temperature may rise above 46°C in the eastern parts of the district in the summer season. Night temperature in June is slightly higher than in May. Day temperature decrease considerably with the onset of south-west monsoon in the month of June. The monsoon season is pleasant. From beginning of October, the day temperature increase by 2 to 3°C. on the average in October as the south-west monsoon withdraws. However, there is a progressive decrease in night temperature after September.

Temperature decreases rapidly from November onwards. December is the coldest month with the mean daily minimum temperature 11.3°C at Malegaon and 10.2°C at Nasik. The mean daily maximum temperature in this month is 29.5°C at Malegaon and 28.3°C at Nasik. Table 2 shows the temperature of last five years in district.

Table 2: Minimum and Maximum Temperature in °C of Selected Sources in Nasik District (1996 to 2001)

Year	*Nasik*		*Ozar*		*Malegaon*	
	Maximum	*Minimum*	*Maximum*	*Minimum*	*Maximum*	*Minimum*
1996–1997	NA	NA	35.4	9.0	NA	24.3
1997–1998	NA	NA	35.4	9.0	NA	24.3
1998–1999	NA	NA	31.8	17.37	30.9	19.59
1999–2000	NA	NA	31.8	16.3	33.5	19.0
2000–2001	NA	NA	39.1	.8.8	41.9	8.7

NA: Not Available.

Source: Meteorological Department, Kulaba, Mumbai.

Humidity and Cloudiness

The air in humid during monsoon season, while it is dry in post-monsoon, winter and summer season. The summer season is the driest part of the year with relative humidity between 20 per cent and 25 per cent only in the afternoons. The sky is heavily clouded during the monsoon season, whereas it is mostly clear or lightly clouded during rest of the year.

Winds

Winds are usually light to moderate with some strengthening in wind force during the latter part of summer season and in the south-west monsoon season. Winds are south-westerly or westerly in the south-west monsoon season. Winds are light in the post-monsoon season and are variable in directions in the mornings and north-easterly or easterly in the afternoons. Winds blow from directions between south-west and north- west in the morning especially in the cold season. They blow between north and east in the afternoon. Winds are from directions between south-west north-west in the hot season.

Vegetation

Although the present study is mainly concentrated on ethnobotanical aspects of plants, it is incomplete without a say about the vegetation and its type in the areas surveyed. The richness of vegetation have attracted many botanists in pre and post-independence period. The diverse content of floristic elements are primarily due to its physiography and climatic conditions, as also courses of several tributaries of the main rivers and its soil texture. The vegetation of Nasik district may be broadly classified in following three main types: (I) Moist Tropical Forests, (II) Dry Tropical forests, (III) Western Sub-tropical Hill Forests.

Moist Tropical Forests

They are subdivided into: (i) West Coast Semi-evergreen Forests, (ii) South Indian Moist Deciduous Forests.

West Coast Semi-evergreen Forests

The forest block Kalmusta in the south and Varasvihir in the north belongs to this type. The density is mainly due to *Mangifera Syzygium*. In the lower valleys tree growth is more akin to the moist deciduous type of forests with a few evergreen species.

South Indian Moist Deciduous Forests

Dominants are mainly deciduous but sub-dominants and lower storeys are mostly evergreen. The canopy in dense and even. This type is further divided into three sub-types namely viz., (a) Moist Teak Forests, (b) Southern Moist Mixed Deciduous Forests, (c) Southern Secondary Moist Mixed Deciduous Forests.

Moist Teak Forests

These are high forest restricted mainly to hilly terrain. Teak is dominant species through out. The rest of the canopy is covered by its miscellaneous associates. The entire top storey consists of deciduous species, whereas lower storey contains a few semi evergreen species. Bamboos are characteristics of the areas.

Southern Moist Mixed Deciduous Forests

The evergreen species is usually larger than in case of teak bearing forest, except for the moist subtype. Teak is observed occasionally.

Southern Secondary Moist Mixed Deciduous Forests

This is restricted to some of the high plateaus. The general composition is similar to that of moist deciduous mixed type of forest but with a higher frequency of semi-evergreen species. It is therefore also known as semi-evergreen forest.

Dry Tropical Forests

This can be further subdivided into (a) Southern tropical dry decidous forest, (b) Southern tropical thorn forest.

Southern Tropical Dry Deciduous Forests

Trees are usually stunded and these forests are situated in moderate rainfall zone. These are found near Khumbala in Nasik range, Wani, and Dindori. This type is also divided into: (i) Dry teak forest, (ii) South dry mixed deciduous forest, (iii) *Hardwickia* forest. The dry teak forest are found on western side of Dindori, north eastern side of Nasik and are restricted to flat and undulating areas. The Southern dry mixed deciduous forest are found. In central part of Nasik district these forests are open and under stalked. The Hardwikia forests occur in Malegaon and Nandgaon ranges where the rainfall is low. The forests are mostly under stalked and blank.

Southern Tropical Thorn Forests

These forest are found in Yeola, Chandwad, Nasik, Sinnar ranges and southern part of Malegaon where rainfall is low. The forests are blank. The grasses are stunted. These forests are scattered and surrounded by cultivation on all sides.

Western Sub-Tropical Hill Forests

These are located in Anjaneri, Trimbak and Igatpuri ranges where rainfall is heavy. It contains semi evergreen and evergreen species. They occur in narrow strips on gentle slopes. This type is divided into two: (a) *Mangifera Syzygium* sub-type which is found in plateau portions. There is a predominance of *Mangifera indica* and *Syzygium cumunii*. (b) *Mangifera Memecylon* subtype is found in Ambevadi and Kurangvadi ranges. The evergreen species of *Mangifera* and *Memecylon* are in high percentage. They are found in narrow strips.

History

The historical background of the district is interesting. The district was occupied in the Early Stone Age tools of trap rocks characteristic of the age *e.g.*, cleavers, scrappers and hand-axes collected from Gangavadi near Nasik. The chalcolithic age is supported by microlitch-trapeze, lunates, and two edged blades of churt and chalcedony in association with orange coloured pottery painted with red or black bands. The literary sources shade light on the history of the region. The whole region was covered by thick jungles, when the Aryans, penetrated this region of The Deccan. Agastya, the celebrated sage, was the first Aryan who crossed the Vindya mountains and fixed his residence on the bank of Godavari. This event is commemorated by the mythical story. Agastya asked Vindhya mountains to remain in the bending posture until he returns from the south, which he never did. The cluster of heritages on the Godavari bank was established by him, knows as Janashana. Panchavati was situated on the fringe of the great forest Dandakaranya, the story of which is narrated in the great epic Ramayana. Danda or Dandaka the grandson of Manu, ruled the country with his capital at Madhumanta. The sage Bhargava cursed him for his misdid. Consequently, the country was turned into Dandakaranya. Rama, Lakshmana, and Sita met Agastya on the bank of Godavari. He presented a bow and two quivers. They were advised by him to settle down at Panchvati, a place named from the five great Banyan trees which grew there. Even now there are some caves which go by the name of Sita gumpha.

The Madmeshwar temple is situated on the banks of Godavari. It is believed that when sita asked for the golden deer, disguised Marich, Rama went after it and killed at the place of Madmeshwar temple.

Katyayana's Varttika (Sirca 250 B.C.) implies the name Nasikya of a city (Nagara) which was evidently the same as modern Nasik. Govardhana which still retains its ancient name lies few miles from Nasik.

The origin of the name Nasik is interpreted in two ways. The town is seated on nine peaks or Navshikhara and hence the name. It is said that Lakshmana, the brother of Rama, cut off the nose (nasika) of Shurpanakha, the sister of the demon king Ravana.

In historical times, this region has been ruled by Maurya's, Shungas, Satavahanas, Abhiras, Traikutakas, Vakatakas, Chalukya's, Rashtrakutas and Yadavas. In the last quarter of

12th century A.D., the period of Yadava Supremacy ended with the invasions of muslims from the north under Ala-ud-din Kilji and Malik Kafur. Muslims ruled the region till 1747. It was known by the name Gulshanabad in this period. It was then ruled by Maratha rulers which ended by British rule.

Plant Based Industries

The importance of industrial development cannot be minimized particularly for a region. It affords employment opportunities and leads to better the standard of people.

Nasik district is predominantly an agricultural district with 73.54 per cent of working population. The general pattern of non-agricultural employment leans heavily on village and cottage industries. The industries can be classified into two types *viz.*, agriculture based industry and industries based on forest materials. Although the economy of this district is predominantly agrarim, in recent years there has been a great development in the industrial sector in large-scale, small-scale and cottage industries. The forests in the district are fairly rich which provide the bases for variety of industries. These are: (i) Wood working industries like saw mills, furniture, packing cases, bullock-carts etc. (ii) Bamboo and grass for paper, (iii) Leaves of Tendu for bidi wrappings (iv) Broom making etc. The district being a predominantly agricultural harbours a variety of cultivated plant species. Many industries are set up for primary processing of crops. These convert them into proper forms for consumptions. *e.g.*, oil industry, flour milling, dehusking of paddy and pulses, ginning of cotton, sugar industry. industry based on grape products etc.

Crop Plants

The district is inhabited by agricultural communities. They occupy especially the plains of the districts as large number of plants are cultivated as crops. There are two seasons as in other parts of the state in which different crops are harvested. They are "Kharif" and "Rabi" an early or rain crop is known as "Kharif", while the crop of cold season is called "Rabi". The important Kharif crops are Rice, Bajara, Jawar, Groundnut, Tur, Cotton, Ragi, etc. whereas crops of Rabi season are wheat, Gram, Varai (common millet) etc. The perennial crops like sugar-cane, grapes and pomegranate are also found in the plains of the district. Following is the list of these economically important cultivated plants:

Cereals and Millets

Eleusine coracana Gaertn., *Oryza sativa* Linn., *Panicum miliaceum* Linn., *Pennisetum americanum* Schum. *Sorghum bicolor* (Linn.) Moench.

Pulses

Cajanus cajan (L.) Millsp., *Cicer arientinum* L., *Lablab purpureus* (L.) Sw., *Pisum sativum* L., *Vigna aconitifolia* (Jacq.) Marechal, *Vigna mungo* (Linn.) Hepper, *Vigna radiata* (L.), Wilczek var.*radiata*, *Vigna unguiculata* (L.) Walp., Lens culinaris Medic.

Vegetable

Abelmoschus esculentus (L.) Moench, *Allium cepa* L., *Brassica oleracea* L. var.*capitata* L., *Brassica oleracea* L. var.*botrytis*, *Cucumis sativus* L., *Colocasia esculenta* (L.) Schott, *Coccinia grandis* (L.) Voigt., *Cucurbita maxima* Duch., *Daucus carota* L., *Ipomoea batatas* L., *Lagenaria leucantha* (Roxb.) Rusby., *Luffa acutangula* Roxb., *Luffa cylindrica* (L.) M.J.Roem., *Lycopersicon lycopersicum* (L.) Sant., *Momordica charantia* L., *Momordica dioica* Roxb., *Moringa oleifera* Lam., *Raphanus sativus* L., *Solanum melongena* Linn., *Trigonella foenum-graecum* Linn., *Spinacea oleracea* L., *Sesbania grandiflora* (L.) Pers., and few others.

Fruits

Carica papaya L., *Citrus sinensis (Linn.)* Osbeck, *Phyllanthus emblica*, *Mangifera indica* L., *Musa paradisiaca* L., *Psidium guajava* L., *Punica granatum* Linn., *Vitis vinifera* Linn., *Ziziphus mauritiana* Lam.

Fibres

Crotalaria juncea L., *Gossypium herbaceum* Linn. var.*acerifolium* (Guill. & Perr.) Chev., *Gossypium arboreum* L., *Hibiscus cannabinus* L.

Oil Yielding Seed Plants

Arachis hypogea L., *Azadirachta indica* Juss., *Brassica niger* (Linn.) Koch., *Carthamus tinctorius* Linn., *Guizotia abyssinica* Cass., *Ricinus communis* L., *Sesamum indicum* Linn., etc.

Spices and Condiments

Allium cepa L., *Allium sativum* L., *Capsicum annuum* L., *Coriandrum sativum* L., *Murraya koenigii* (L.) Spr. etc.

Ethnology of Tribes in the District

Apart from non-tribal people there are many tribes inhabiting in Nasik district. They are Mahadeo-Kolis, Bhils, Kokani (Kokana kunbis), Vanjaris, Thakurs, Varlis, Vadars, Kaikadis, Katkaris, Vaidus, Kangaris, Bhamta's (Uchlas) and Berads.

Bhils

Bhils are said to be migrants from the adjacent Dang region of the state of Gujarat. They have settled in talukas like Kalwan, Satana (Baglana) and Malegaon and rarely in some other parts of the district. They are strong, dark, active and hardy with high chick bones and wide nostrils. In some cases, they show coarse features. They are known for their bravery, carefree desposition and excellance at archery. They speak Bhilly dialect which is mixture of Marathi, Gujarathi and Hindi. Their dress is simple. A male wears traditionally a turban (now-a-days replaced by a cap), loin cloth or short dhoti as a waist cloth and Sadara (Shirt). Their women wear coarse Sarees and Choli (Blause).

Kokanis

They speak Kokani language which is mixture of Marathi, Hindi, Arabic and with some Konkani peculiarities. The Kokanis have regular and clear-cut features with generally an expression of keenness and intelligence. They are tall and muscular though spare with fair, ruddy or clear olive skins and black or brown eyes. The younger men wear the hair hanging to the low of the ear, whereas the older shave the head. Both young and old wear full curly beards. The gents have common Muslim dress, whereas ladies wear the Maratha robe and bodice. However, their ornaments are the same as those worn by Muslim women. They are clean in their habits, hard working, sober and fond of amusement, charitable and hospitable. They are sunnis of the Shafis school.

There are also Hindu Kokanis inhabiting the district. They are imigrants from adjacent Thana district of Maharashtra. They regard themselves Konkan Kunbis. They live in the traps of west Aravle land. They occasionally go to their native homes in Konkan region of Maharashtra to renew their connections.

Katkaris

Katkaris are also immigrants from the Konkan region. They are usually found within the limits of Sahyadris. They are the poorest of Nasik forest tribes. Their women wear a ragged shoulder cloth across the breast. They speak corrupt Marathi mixed with Gujarathi words. One branch of the tribe Dhor Katkaris eat beef. They are originally catechu-makers. Their God's are Chaide and Mhasoba. They worship ghosts and demons. They have no priests and themselves officiate at marriage ceremonies.

Mahadev Kolis

They are found on along the Western Ghats. They are a fine looking race. The most civilized and settled of all hilly tribes. They grow hill grains, pulses, oil seeds and rice. They have taken firmly to tillage and husbandry. The rites and customs are same at Kunbi weddings in the district.

Thakurs: Thakurs are said to have come from upper India in search of military services. They are tall, strongly made and dark brown. The men let their hair grow and wear the beard. They speak Hindustani (Hindi) at home but also know Marathi well. The men dress like Maratha's and women in 'Pardeshi' fashion with a robe, Lahanga (Phadki) and a bodice (Choli) and when they go out a white sheet (Chadari). They are clean in their habits, soldier-like, hot tempered, hard working and orderly. Their ancestors are said to have served under the 'Peshwas" as soldiers. They eat animal food but do not touch foal or cow and buffaloe's beef. They do not eat onions or drink. The men wear a sacred thread like Bramhans. Marriages are perform in North Indian Style. The bride and bridegroom are required to walk seven times round a pillar fixed in the marriage booth. They worship Devi, Khandoba, Mahadev and Rama. Their priests are Kanoja Brahmans who officiate at their marriages.

Varlis: Varlis are uplanders found in Peth taluka and on the Western Ghats. Their name seems to appear in Varalatta (The most Northerly but one of the seven Hindu Konkan. Like Thakurs, they live part of year on grains they raise and for the rest almost entirely on forest products.

Previous Ethnobotanical Explorations in Maharashtra

Ethnobotany was initiated by Dr.E.K.Janki Ammal (1897–1984). She studied subsistence of food plants of certain aborigines of south India. She opened up an "Ethnobotanical Section" of Botanical Survey of India at Allahabad in 1960. Dr.S.K.Jain searched intensively among the tribals of central India. He triggered off ethnobotanical activity in India by impressing on the scientists of different disciplines.

As far as state of Maharashtra is concerned, it was Late Dr.V.D.Vartak, his associates and students extended ethnobotanical observations in the western part of Maharashtra. Their mission is still continued in this region. The data on ethnobotany in Maharashtra is largely available through their efforts (*cf.* Ghate, 1998; Ghate and Vartak, 1996; Ghate, Agate and Vartak, 1988; Ghate, Vartak and Kumbhojkar, 1990; Gadgil and Vartak, 1973, 1974, 1976, 1981; Gaikwad, 1988; Godbole, 1934, 1994, 1996; Gunjatkar and Vartak, 1982, 1983; Kulkarni and Kumbhojkar, 1992a-c, 1993, 1995, 1996, 1999, 2000; Kulkarni, Kulkarni and Kumbhojkar, 1996; Kulkarni *et al.* 1991; Kulkarni, Agate and Kumbhojkar, 2003; Kulkarni, Kumbhojkar and Khedkar, 1998; Kumar, Kulkarni and Singh, 1980; Kulkarni, Kumbhojkar and Nipunge, 1990; Kumbhojkar, Kulkarni and Upadhye, 1991; Kulkarni, Kumbhojkar and Upadhye, 1995; Kumbhojkar, Upadhye and Kulkarni, 1996; Kumbhojkar and Kulkarni, 1998; Nilengaonkar, Vartak and Chitre, 1985; Nipunge, Kumbhojkar and Vartak, 1988; Nipunge, Vartak and Kumbhojkar, 1992; Nipunge, Kulkarni and Vartak, 1993; Upadhye and Kumbhojkar and Vartak, 1988, 1992, 1996, 1998; Upadhye, Vartak and Kumbhojkar, 1994; Upadhye, Kumbhojkar and Kulkarni, 1997a-b, 1999; Vartak, 1957, 1959a-c, 1970, 1971, 1980, 1981-a-b, 1982, 1990; Vartak and Gadgil, 1980, 1981, 1990; Vartak and Kulkarni, 1987; Vartak and Mandavgane, 1981; Vartak, Kumbhojkar and Ghate, 1985; Vartak, Kumbhojkar and Nipunge, 1987; Vartak and Ghate, 1990, 1994).

Some others, as a part of their floristic studies, also added information from this region (*cf.* Almeida and Almeida, 1987; Caius and Radha, 1939; Jayanand Tosh, 1981; 1996; Jayanand Tosh, Vartak and Kumbhojkar, 1959; Ryan, 1903; Tribhuvan and Peter, 1992; Vedprakash and Mehrotra, 1987). Scientists of Botanical Survey of India, Western Zone, also added few publications as a part of their floristic study in the state. (*cf*. Anand Kumar, Kulkarni and Singh, 1981; Ansari and Rao, 1973; Ansari and Nayar, 1978; Ethoven, 1930; Janardhan, 1961, 1963; Kamble and Pradhan, 1980; Lisboa, 1887; Mudaliar, Kulkarni and Sharma, 1987; Rao, 1960; Puri, 1959, 1960; Santapau, 1956; Shah, 1984; Shah, Yadav and Badrinath, 1983; Sharma and Malhotra, 1984). Few research papers have also been added over years (*cf*. Chandras, 1966; Datar and Vartak, 1975; Deokule and Magdum, 1992; Kirtikar, 1906; Sensarma, 1984; Kurian and Bhanu, 1980; Kurian and Tribhuvan, 1990; Shekhawat and Sunil Anand, 1984; Tarafder, 1984; Tekale, 2002).

Vidarbha region of the state of Maharashtra has made a headway in this line (*cf*. Badhe and Sharma, 1981; Badhe and Pandey, 1988; Badhe, Sharma and Pandey, 1997; Bhogaonkar and Devarkar, 2001a-b; Chute and Tiwari, 1999; Das and Misra, 1988; Deshmukh and Dhore, 1994; Deshmukh, Muratkar and Rothe, 1999; Giri, 1994; Malhotra and Moorthy, 1973; Mirashi and Salpekar, 1975; Padhye, Deshmukh and Tiwari, 1992; Tiwari and Padhye, 1993).

From Konkan region of the state are added few publications (*cf*. Ghate and Sane 1994; Kothari and Moorthy, 1987, 1996; Kulkarni, 1968; Mokat and Deokule, 2004).

Kolhapur district of Maharashtra also has been surveyed ethnobotanically (*cf*. Khandekar and Badve, 1968; Mahajan and Divan, 1968; Magdum and Gajage, 2004; Majgaonkar and Phadke, 1999; Sharma and Kulkarni, 1980; Upadhye, Kumbhojkar and Vartak, 1986).

Recently Marathwada region of the state has received some attention of botanists (*cf*. Bhutkar, 2002; Gupta, Hussain and Imam, 1997; Korekar, and Papdiwal, 2002; Naser Rafiuddin and Vaikos, 2002; Rathor, Zare, Kagne and Chavan, 2002; Rohidas and Bankar, 2002; Wangikar, 2002).

Khandesh region, especially Dhule and Nandurbar Districts have been investigated through the efforts of Patil, Borse and Bhamare, 1990; Bhamare, 1995, 1998a-b; Patil, 1986, 1998a-b; Patil and Bhamare, 1996; Yadav and Bhamare, 1989. As far as Jalgaon and Nasik Districts are concerned, Karnik, 1966; Kshirsagar and Patil, 2002; Lakshminarasimhan and Sharma, 1990; Shisode and Patil 1993; Pawar and Patil, 1999, 2000a, 2001, 2003, 2004a-c; Pawar, Patil and Patil, 2004; Patil and Patil, 2000a-c, 2001, 2002, 2003a-d;

Patil, Pawar and Patil, 2002, 2003, 2004; Patil, Patil and Pawar, 2004a-b; Sharma and Lakshminarasimhan, 1986; Tayade and Patil, 2003a-c published few research papers.

The survey of pertinent literature suggests that Nasik District has largely remained untapped ethnobotanically. The present effort undertaken, therefore, appears amply justified. This effort has generated vast indigenous knowledge and will be very much relevant to the conservation of biodiversity and sustainable use of plant resource of this district.

Earlier Botanical Explorations in the District

In pre-independence period, a large number of botanists visited the district for botanical explorations. These are Graham, Dalzell and Gibson, Nainare, Lisboa, Woodrow, Cooke, Edgeworth, Talbot, Kanitkar, Bhiwa, Wadekar, and few others. Some other workers from B.S.I., Pune, like Puri, Jain, Vasavda, Gangurde, A.S.Rao, Rolla Rao, Cherian and Pataskar botanised in the district. Very recently, Lakshminarsimhan and Sharma (1991) surveyed the district floristically and published their floristic account, covering as many as 943 species belonging to 108 families. They obviously paid cursory attention to the economic significance of the plant wealth of the region. They gave some information on ethnobotanical line as a part of their routine floristic studies. The district has not been studied ethnobotanically as an exclusive topic of research to date.

Methodology

Intensive field work was undertaken since June 1997 covering all the seasons so as to collect detail information on plant species found useful in ethnomedicine as well as for the other local uses of the plants occurring in Nasik District of Maharashtra. During ethnobotanical surveys all tribal localities, adjoining forest areas and the plains of the district were visited. Routine methods of botanical collections and techniques of herbarium preparations were followed as suggested by Jain and Rao (1977). Plants have been collected in flowering and fruiting stages and serially tagged with collection numbers. They are denoted after the investigator's name (MVP). Observations were made of the plant species with respect to their location, habit, habitat and other field characters. The plant specimens were identified using district, regional and state Flora's. Following were especially consulted: (i) Flora of Nasik

District by Lakshminarsimhan and Sharma (1991), (ii) The Flora of the Presidency of Bombay by Cooke (1958, BSI Repr.), (iii) Flora of Marathwada by Naik (1998), (iv) Flora of Maharashtra state: Monocotyledons by Sharma, Karthikeyan and Singh (1996), (v) Flora of Maharashtra State: Dicotyledons Vol.I by Singh and Karthikeyan (2000), (vi) Flora of Maharashtra: Dicotyledons Vol.II by Singh and Kartikeyan (2001) (vii) Flora of Dhule and Nandurbar District (Maharashtra) by Patil (2003).

The herbarium specimens and parts collected have been housed in the Herbarium. Department of Botany, Gangamai Education Trust's Arts, Commerce and Science College, Nagaon, Dhule District, Maharashtra.

While collecting information on various local uses of plants, the approaches and methodologies have been followed as suggested by Jain (1987–1989). The witch-doctors, Buvas, Bhagats, and other learned informants like old men and women or heads of hamlets and villages etc. were interviewed repeatedly to collect the first-hand information. Information was gathered regarding plants or plant parts used, recipe of medicines, doses, type of ailments or diseases and plant part used for other miscellaneous purposes. Plants related with worships, beliefs, birth, deaths, marriages, festivals etc. are also noted. Vernacular names of plants appeared interesting and therefore have been also noted while botanising in the district. Ethnological information of the main tribes have been noted as available from published literature. Few photographs of tribals, plants or their parts collected during field work are presented to authenticate the information accrued.

Ethnobotanical data gathered from the district has been compared with the classical literature like: (i) The Wealth of India: A Dictionary of Indian Raw Materials and Natural Products. Vol.I to 11 (1948–1976), (ii) A Dictionary of Economic Products Vol. I-VI by Watt, G. (1889–1893), (iii) The Useful Plants of India by S.P.Ambasta (1986) and (iv) Dictionary of Indian Folk Medicine and Ethnobotany by S.K.Jain (1991) and recent publicatons. This data is considered under notes from classics.

Readers or users of this contribution are requested to note that the information abstracted from classical literature is usually documented referring old names. However, the present investigators have selected their new/valid names for such taxa.

Ethnobotanical information gathered from Nasik District is presented under three headings under Ethnobotanical Notes *viz.*, Medicine (*i.e.*, human medicine), Veterinary medicine and the rest others uses under Miscellaneous. In case of vernacular plant names, however, different pattern has been followed to suit the data accrued and hence considered under a separate chapter.

In case of ethnobotanical information from Nasik District, the recent botanical name, family name and local name are given for each species. The plant species have been arranged alphabetically. The families are arranged according to Bentham and Hooker's system of classification with slight modifications in the treatment of families adopted in recent floras. Vouchers specimens are housed in the Herbarium of Gangamai Education Trust's Arts, Commerce and Science College, Nagaon, Dist. Dhule. Single aster (*) in the ethnobotanical enumeration stands for additional use or application of the plant species. Double asters (**) are given for those species which are noted useful for the first time.

Chapter 2
Ethnobotanical Enumerations

Abelmoschus esculentus (Linn.) Moench.

Family: Malvaceae, Local Name: Ranbhendi

Ethnobotanical Notes

Medicine

1. A spoonful decoction of roots is given twice a day to treat burning sensation and yellow urination.
2. Roots are chewed to control dysentery.
3. Roots paste is applied on injuries.

Miscellaneous

1. Fruits are cooked as vegetable and consumed.

Notes from Classics

Leaves and fruits are employed externally as emollient poultice. Muscilage from fruits and seeds is considered as demulcent, used on gonorrhoea and irritation of genito-urinary system. Young pods are eaten by natives as a treatment for spermatorrhoea. Unripe fruits are favourite vegetable. Seeds are used as substitute for pearl, barley and coffee, eaten with roasted bread. Bast yields strong and useful fibres.

(Watt, 1889–1893)

Leaves are employed externally as emollient poultice. Infusion of roots is reported useful on syphilis. Seeds are used as a substitute for coffee, also used in curries and chutnies. Seed oil is bleached and deoderised, used as a substitute for edible oil. Tender leaves are boiled and eaten as spinach. Tender pods are cooked as vegetable. Muscilaginory extract of green stem is employed for clarifying sugarcane juice in manufacture of gur, also reported as used for paper sizing. Stem fibre is used for paper and card board manufacture. Leaves are used cattle fodder. Stems are used as fuel.

(Anonymous, 1948–1976)

Immature capsules are emollient, demulcent and diuretic. Seeds are stimulant, cordial and antispasmodic. Ripe seeds are roasted for their use in coffee, also used in curries and chutney. Seeds are rich in protein, powdered and mixed with maize flour. Tender pods are used as vegetable, also used for thickening of soups and gravies. Tender leaves are used as spinach.

(Ambasta, 1986)

Fruits are used as abortifacient. Fruits are cooked and eaten as vegetable.

(Jain, 1991)

Critical Notes

Roots are medicinal to treat burning sensation during urination and also for yellow urination. Plant species is exotic, native of tropical America.

**Abelmoschus tetraphyllus* (Roxb. ex Horn.) Borssum

Family: Malvaceae, Local Name: Garabhendi

Ethnobotanical Notes

Medicine

1. Roots are chewed to control dysentery.
2. Extract of stem bark alongwith bark of *Oroxylum indicum* (L.) Vent. is admistered to treat Leucorrhoea.

Notes from Classics

Roots are used on pneumonia. Roots are edible.

(Jain, 1991)

Critical Notes

Chewing of roots to control dysentery and leaves of bark against leucorrhoea are new application.

**Abrus precatorius* L.

Family: Papilionaceae, Local Name: Gunj

Ethnobotanical Notes

Medicine

1. Roots pieces are kept in water. This infusion about 15–20 ml thrice a day for two days is orally given to ladies to control white discharge.
2. This infusion also given against cough and asthma.
3. Roots are chewed twice a day for 8–10 days to treat kidney stone.

Notes from Classics

Root is a substitute for liquorice. Infusion of roots is used for abortion in Hoshiarpur (Punjab). Cold infusion is used in Leucorrhoea. Leaves are heated on fire and applied on swellings. Leaves steeped in mustard oil are applied over seat of pain in rheumatism. Seeds are used as purgative but in large doses are acid and poisonous. Seeds boiled with milk are said to have powerful tonic action on nervous system. Seeds are pounded, mixed up with mercury, sulphur, neem seeds, hemp leaves, and cotton seeds etc.

are externally used on skin diseases. Seeds are used by Sweepers and Chamors to poison cattle. Red seeds are used by Indian Goldsmiths as weight, each weighing 1.75 grains. They are also used in ornaments and necklaces.

(Watt, 1889–1893)

Decoction of roots and leaves is used on cold, cough and colics. Leaves are chewed as remedy for hoarsness. Infusion of seeds is poisonous if applied on conjunctiva. Poultice of seeds have been used as a suppository to bring about abortion. Seeds are used for cattle poisoning and homicidal purposes. Seeds are used as beads for necklaces and ornaments.

(Anonymous, 1948–1976)

Roots are abortifacient, anthelmentic, used in bronchitis, cold, febrifuge, purification of blood, leucoderma, rheumatism, skin diseases, scabies, swellings, sore throat, urine complaints and on wounds. Fruits are used in eye complaints. Seeds are aphrodisiac, antifertility, eczema, paralysis and pneumonia. Seeds are purgative and tonic, used to treat stomach complaints and sciatica. Plant is useful in childbirth, convulsion, diarrhoea, inflammation of gums, leucorrhoea, menstrual complaint, muscular pains, urine complaints and bone fracture. Roots are used against scorpion sting. Leaves are used as veterinary medicine on wounds and as masticatory. Seeds are used as weights by Goldsmiths. Seeds are used in necklaces and ornaments.

(Jain, 1991)

Critical Notes

Use of roots to treat kidney stones is being reported additionally.

Acacia auriculiformis A.Cunn. ex Benth.

Family: Mimosaceae, Local Name: Akashya

Ethnobotanical Notes

Veterinary Medicine

1. Fruits are boiled and vapours are directed towards crop fields as insecticide.
2. Fruits are boiled and vapours are allowed on septics to repel worms from wounds of cattle.

Miscellaneous

1. Fruits are crushed and used as fish poison.

Notes from Classics

Bark contains tannin used for tanning. Seeds yield fatty oil. Plant is useful as sand binder.

(Ambasta, 1986)

Critical Notes

Fruits are probably very poisonous. Therefore, they have been used as insecticidal, piscicidal, on septic and repeling worms from wounds of pet animals. All these are new applications of this taxon. It should be verified on scientific modern lines so as to find out an alternative drug or source of chemical for aforesaid purposes. The plant species exotic, native of Tropical Australia. It is introduced tree species as avenue.

**Acacia catechu* (L.*f.*) Willd.

Family: Mimosaceae, Local Name: Khair

Ethnobotanical Notes

Medicine

1. Paste of fresh leaves is applied on injury.

Miscellaneous

1. Stem pieces are placed in rice crop at random distances to prevent the crop from yellow leaf disease.
2. Wood is used for preparing cart wheels.
3. A tassel is made from its wood to prepare and clean food grains.

Notes from Classics

Kath is obtained from bark. Small piece is kept in mouth, and allowed to dissolve as a remedy in relaxation of 'Uvala' and irritation of fauces and trouble some cough. It is also useful in mouth hoarseness. Decoction of catechu is used to wash inflammed parts and ulcers. Catechu alongwith nutmeg, camphor and betel nut, are made into balls. It is also used in ptyalism, ulceration, spongyness of gums and hypertrophy of tonsils. Powder of catechu mixed with water is used in conjunctivitis. Powder if mixed with egg yolk commonly applied to cancers in east Africa. Power of bark is mixed with ghee and ointment obtained is applied on Syphilis and burns. Bark powder mixed with aromatics is used in melancholia. Bark powder is used as remedy for intermittent fever and scurvy. Infusion of 'Katha' is applied for sore nipples as preventive against ill effects of nursing wash is given for some weeks before confinement. Mixture of catechu and myrrh called Kathbol, generally given to women after confinement as a tonic to promote secretion of milk. Infusion of 'Kath' is used by 'Hakims' for the treatment of abortion, it is useful for women who are barren but are desiours of having offspring. Resinous extract is a powerful, astringent, internally useful in diarrhoea with pyrosis. An injection of aqueous solution of 'Katha' is often used in leuccorrhoea and as tonic. By Hindu physicians 'Katha' is externally and internally applied on skin diseases. 'Katha' is pasted on betel leaf and chewed. Timber is used for rice pestles, oil and sugar-cane crushers etc. Wood is more valued fuel for Goldsmiths.

(Watt, 1889–1893)

'Katha' is regarded as astringent, cooling and digestive, useful in relaxed condition of throat, mouth and gums, also used as a remedy for cough and to control diarrhoea.

(Anonymous, 1948–1976)

Uses recorded by Ambasta, (1986) are similar to those given by Watt (*loc.cit.*).

Roots are used on tooth-ache. Stem bark is used in asthma, bleeding from nose, colic, dysentery, chest pains, sore mouth, leucorrhoea and menstruation complaints. Leaves are used to treat gonorrhoea and heomoptysis. Heart wood is inhaled for bronchitis and cancer. Entire plant is used in child-birth.

(Jain, 1991)

Critical Notes

The wood being considerably hard is specially used for cart wheels and tassel. These are still unreported.

Acacia ferruginea DC.

Family: Mimosaceae, Local Name: Vichan

Ethnobotanical Notes

Miscellaneous

1. Wood is especially used for arcs of wheels of bullock-cart and agricultural tools.

Notes from Classics

Bark possesses astringent property. Gum is having similar qualities of gum of *Acacia nilotica* (Linn.) Delile. Timber is used for building carts and other agricultural implements.

(Watt, 1889–1893)

Stem and pods bark possess astringent property. Gum has similar properties of gum of *Acacia nilotica* (Linn.) Delile. Wood is used for cart wheels, poles, beams, and agricultural implements.

(Anonymous, 1948–1976)

Wood is used for cart wheels, post, beams and agricultural implements.

(Ambasta, 1986)

Stem bark is medicinally astringent. Wood oil is applied on scabies.

(Jain, 1991)

Acacia leucophloea Willd.

Family: Mimosaceae, Local Name: Hiwar

Ethnobotanical Notes

Medicine

1. Roots are rubbed on stone with water and is applied, swellings and plummet.

Miscellaneous

1. Wood is used as fuel.
2. Leaves are used as goat fodder.

3. Green fruits are used as fodder especially for goats.
4. Wood is used for making handles of axe and other agricultural implements.

Notes from Classics

Stem bark is astringent. Stem bark is ground and mixed with flour in times of scarcity. Fruits are largely used as fodder in Punjab. Wood is used as fuel.

(Watt, 1889–1893)

Stem bark yields fibres used for fishing nets, and coarse cordage. It is also used for tanning. Bark is used in preparation of spirits from sugar and palm juice hence knows as distiller's Acacia or 'Sharab ki Kikar'. Timber is used for agricultural implements, carts, wheels and turnery, also used as fuel.

(Anonymous, 1948–1976)

Ambasta (1986) recorded used similar to those by Anonymous (*loc.cit.*).

Roots are abortifacient. Stem bark is astringent, used on ulcers. Wood symbolically charms.

(Jain, 1991)

**Acacia nilotica* (Linn.) Delile Subsp. *indica* (Benth.) Brenan

Family: Mimosaceae, Local Name: Babul

Ethnobotanical Notes

Medicine

1. Apical buds are crushed in water and juice is obtained. About 10–15 ml of it is administered at night to treat cough.
2. Leaf paste is applied on swelling or inflamed body part.

Miscellaneous

1. Wood is used for hut and house construction.
2. Wood is also used for making plough and cultivators and handles of agricultural implements.

3. It is employed for fuel.
4. Dried branches are used for fencing crop fields and farmyards.
5. Leaves and green pods are fed to goats.
6. Stem pieces are used as tooth-brush.

Notes from Classics

Slender stem is used for cleaning teeth. Decoction of bark is used as astringent gargle for sore throat, stomatitis and ulcers of mouth. It is also useful in chronic dysentery and diarrhoea. It also has been valuable remedy in prolapsus, ani' as an external applicant in leucorrhoea, and mennorhagia, recommended as a poultice for ulcers attended with serious discharge. Decoction of bark is used in caries of teeth as mouth wash. Bark and seeds are burnt and this powder is used as tooth powder. Leaf juice is dropped into eyes for epithora and conjunctivitis. Pulp of tender leaves is given in diarrhoea. Tender leaves of this plant and leaves of pomegranate are beaten into pulp and used in gonnorrhoea. Tender leaves fried in ghee, are used as poultice over eyelids to remake chronic congestion of conjuctiva. Tender leaves are bruised, mixed with human milk also used as a poultice on conjunctivitis. Juice mixed with milk is dropped in eyes. Tender legumes are useful in diarrhoea and dysentery. Gum is administered as tonic to a recently delivered woman. Gum is largely used in dysentery and diarrhoea, diabetes mellitus. Bark yields dye which is used by tanners in leather industry. Leaves are also used for tanning and dyeing. Leaves are an important fodder in times of scarcity. Gum is used by calico-printers and for other industrial purposes. Stem bark powder is dusted over sores and ulcers on the lips of horses. Stem bark powder is used as antidote against snake-bite in Punjab. Green leaves, tender shoots and pods are given as fodder to sheep, cattle, goats and camels. Wood is used for carts, agricultural implements and tool handles. Bark is used as a substitute for soap. Green pods are made into ink. Twigs are used for fencing. Bark is largely used in distillation of rum. Wood is used as a popular fire wood.

(Watt, 1889–1893)

Ambasta (1986) and Anonymous (1948–1976) mentioned simlar uses as those in Watt (*loc.cit.*).

Roots are used on skin diseases. Stem is used as tooth-brush. Stem bark is used on bronchitis, diarrhoea, syphilis and burns. Leaves are used against hair-fall, leprosy and also used as tonic. Flowers are used in urinal complaints. Entire plant is useful in diabetes, throat infection and also in ear, eye and liver complaints. Gum is used on cholera as masticatory.

(Jain, 1991)

Critical Notes

Use of apical buds against cough and of leaves as a remedy for body swelling or inflammation are not on classical records. Plant species is exotic, native of North America.

**Acacia polycantha* Willd.

Family: Mimosaceae, Local Name: Ghubeta

Ethnobotanical Notes

Medicine

1. Stem bark paste is applied on swelling caused on account of snake-bite.

Critical Notes

The species is not mentioned in classic literature. Obviously, it is a new report of its application on swelling after snake-bite.

Acalypha malabarica Muell.-Arg.

Family: Euphorbiaceae, Local Name: Khokali

Ethnobotanical Notes

Miscellaneous

1. Plant is used as cattle fodder.

Critical Notes

It is a common weed of cultivated lands. It is removed manually by the farmers and used as such. This appears a proper use of this weedy species.

Adathoda zeylanica Medic.

Family: Acanthaceae, Local Name: Adulsa

Ethnobotanical Notes

Medicine

1. About two spoons of leaf decoction is given twice for two to three days for cough.

Notes from Classics

Roots and leaves are very efficacious remedy for all sorts of coughs. Dried bark when smoked relieves asthmatic fits. It is much benefited as expectorant. Leaves are smoked in asthma. Decoction of leaves is useful in bronchital cattarah. Decoction of leaves is considered efficacious is rheumatic pains and neuralgia, also useful in reducing swellings. Decoction is also used to quench thirst in fever. Leaf juice is useful in diarrhoea and dysentery. Fresh flowers are bound over eyes in opthalmia and are supposed to be antipasmodic. Plant is useful in the treatment of chronic bronchitis and asthma. Leaves are supposed to be high quality manure. Timber of thick stem is used for charcoal of gun powder.

(Watt, 1889–1893)

Root and leaves are antispasmodic and efficacious in cough. Drug is useful as expectorant and mild bronchial antispasmodic. Leaves release spasm. Vasica drug is extracted from plant, used on bronchitis, asthma, fever, jaundice and consumption. Leaves are used as green manure. Leaves yield yellow dye.

(Anonymous, 1948–1976)

Fresh or dried leaves are used on bronchitis and consumption. Leaf juice is also used on diarrhoea, dysentery and glandular tumours. Leaf powder is used on skin affections. Plant is employed as green manure. Wood is used for charcoal and gun powder.

(Ambasta, 1986)

Roots are anthelmintic, antipyretic, antiseptic, antispasmodic and used as expectorant. Roots are useful in cases of child-birth and eczema. Leaves are useful in asthma, bronchitis, cough, malaria, to cure dislocations of joints, rheumatism and swellings. Leaves and flowers are used in cold and cough. Leaves, flowers and roots are used in pulmonary affections. Seeds are used on boils. Whole plant is useful in consumption and veneral complaints. Wood is used as beads for rosaries.

(Jain, 1991)

**Achyranthes aspera* L.

Family: Amaranthaceae, Local Name: Aghirda

Ethnobotanical Notes

Medicine

1. Root juice is dropped on place of scorpion sting. It gives relief from pains.
2. Leaves of these plant and those of *Dolichos lablab* L. are crushed. This juice is applied on body to treat body fever.
3. Entire plant is uprooted on Saturday morning and roots are chewed to remove thorn inserted deep in foot sole.

Notes from Classics

Decoction of entire plant is laxative and promote secretion, used in combination with other medicine in ascites and aṇasaraca, fumed efficacious in the treatment of dropsy, piles, boils and eruptions of skin. Seeds and leaves are emetic, used in hydrophobia and snake-bite. Ash of plant is astringent and diuretic, recommended in the treatment of menorrhagia and diarrhoea. Bengal women used ash of entire plant to produce abortion. The prickly flowers being inserted in uterus, or juice is injected into uterus, labour pains will be set up rapidly. Ash mixed with orpiment externally used in the treatment of ulcer and warts on body. Ash mixed with sesamum oil is used on ear diseases. Plant powder is given to children for colics. Flowering spikes are used as against scorpion bite.

(Watt, 1889–1893)

Paste of spikes or seeds, is used as an application for poisonous insects. Decoction of entire plant is said to be diuretic, occasionally useful on renal dropsies. Young leaves are served as spinach in the Moluccas.

(Anonymous, 1948–1976)

Decoction of herb is diuretic, used in renal dropsies. Young leaves are served as spinach. Ash is rich in potash, and can be used as manure.

(Ambasta, 1986)

Roots are useful against antifertility in women, bleeding in delivery, child-birth, boils, cold, colics, ear complaints, piles, wound and also reduces hunger. Leaves are useful in dropsy, eye complaints, febrifuge, leucoderma and skin diseases. Inflorescence is used for colics. Entire plant is used against bronchitis, dysentery, head-ache, liver complaint, rheumatism, scabies, and whooping cough. Plant is also used in hydrophobia, germicide and antidote on snake-bite, and insect-bite. Seeds are used in renal complaints and also used in tooth-ache. Roots are used as veterinary medicine to expel placenta to ward of ghosts.

(Jain, 1991)

Critical Notes

Leaves are useful for fever. Roots are employed in scorpion sting and to remove thorn. All these did not find place in classical literature yet. The plant species is exotic, native of North Africa.

Adansonia digitata L.

Family: Bombacaceae, Local Name: Gorakhchinch

Ethnobotanical Notes

Veterinary Medicine

1. Fruits are dried and powdered. It is mixed with water and given to animal suffering from dysentery.

Miscellaneous

1. Fruits are hanged in house and shops to prevent from devil's power.

Notes from Classics

Bark and leaves are used as condiment by Negroes. Leaves are eaten by fishermens of Gujarat with their food. Leaf powder 'Lalo' is used to check excessive perspiration. Fruit pulp is good refrigerant in fever. It is used on dysentery in Africa. Fruit pulp with butter milk given on diarrhoea and dysentery in Bombay. Seeds have febrifugal properties. Seed pulp is externally applied on skin diseases. Decoction of seeds is used in bilious head-ache. Stem bark yields strong fibre, used for rope making and also woven into clothes. In India, elephant saddles are made from it. Fibre is used in paper industry. Ash of fruit and bark boiled in oil is used as soap by Negroes. Dried fruits are used as floats by fishermens.

(Watt, 1889–1893)

Leaves are used as prophylactic used against fever during rains and also checks excessive perspiration. Fruit pulp is employed in the treatment of fevers and in dysentery. The fibre is used for rope making and sacking.

(Anonymous, 1948–1976)

Fruit pulp gives relief in bronchial asthma and allergic dermitis. Kernels are eaten, oil extracted from these is used for cooking purpose. Bark yields soft fibre, used for rope and sac making. Wood is used for paper manufacturing, fishing nets, floats, canoes, rafts and trays.

(Ambasta, 1986)

Fruits are used as fishing floats. Seeds are used for polishing pottery. Tree is symbolically sacred.

(Jain, 1991)

Critical Notes

Abstract use of fruits to nullify devil's power is interesting and possibly not reported earlier. It is exotic tree species, native of tropical Africa.

Adiantum philipense Linn.

Family: Adiantaceae, Local Name: Kesari

Ethnobotanical Notes

Miscellaneous

1. Stem pieces are kept in 'Taite' (amulet) and fastened around the neck or arm to prevent from magic power.

Notes from Classics

Rhizomes are prescribed for strangury and in fever due to eplephantasis. Fronds are burnt and applied in oil for itch. Entire plant is used in blood diseases and epileptic fits.

(Ambasta, 1986)

Leaves are used on asthma, atrophy, cough, fever, diarrhoea, bone fracture, pimples and muscle pain etc. Plant is also useful in rabid dog-bite.

(Jain, 1991)

Critical Notes

Use of stem to prevent magic of others is interesting.

**Aegle marmelos* Correa ex Roxb.

Family: Rutaceae, Local Name: Bel

Ethnobotanical Notes

Medicine

1. Scrapings of fruit and coconut *Cocos nucifera* L. in the leaves of *Piliostigma malabaricum* (Roxb.) Bth. and this bidi is smoked for sores or infections of throat.
2. Juice of flowers of this plant and that of *Erythrina stricta* Roxb. is given on cough. Two spoonful a day is administered daily.

Miscellaneous

1. Leaves are offered specially on the idol of Lord Shiva.
2. Ripe fruits are eaten by tribals.

Notes from Classics

Decoction of root bark is used for intermittent fever. It constitutes as an ingradient in (Dasmula) 'ten roots' given on hypochondriasis and palpitation of heart. Leaf juice is used on cattarah and feverishness. Leaves are used as poultice in opthalmia. Unripe fruits are astringent. Fresh ripe fruits are laxative. Roasted or sun dried unripe fruits are cut into slices and are useful for diarrhoea and dysentery. Ripe fruits are sweet, aromatic sherbet is prepared, cooled with ice and given in cases of fibrile affection and dyspepsia. Bel marmalade is taken like jam at breakfast table, for chronic dysentery and diarrhoea. Ripe fruits are used for dyeing and tanning.

(Watt, 1889–1893)

Roots and stem bark are administered in intermittent fever. Unripe fruits are regarded as digestive and stomachic. Ripe fruits are sweet, generally used for preparation of sherbet and marmelade. Stem bark is used as fish-poison. The dried fruits freed from pulp are used as pill boxes for keeping medicines, sacred ashes etc. Gummy mucous substance around seeds is used as adhesive. Leaves are much used as religious offerings in temple of Shiva. Young fruits are substituted by seeds of *Elaeocarpous ganitrus* Roxb. (Rudraksh) for making necklaces. Wood is suitable for making charcoal.

(Anonymous, 1948–1976)

Ambasta (1986) reported uses as given by Anonymous (*loc.cit.*).

Roots are used against palpitation of heart. Stem bark is used in diarrhoea, dysentery and stomach disorders. Roots are useful against mad-dog bite. Roots bark is used as fish poison. Leaves are medicinally used on diabetes, abscesses, back-ache and eye complaints. Leaves are added in bath water and bath is given to person suffering from fever, heat in abdomen, jaundice, vomiting, wounds and cuts. Leaves are used as veterinary medicine. Fruits are used in constipation and as tonic. Plant is useful in abdomen pain, breast pain, cholera, cramps, nausea, night fever, post-natal and puerperal fever. Plant parts are useful in snake-bite. Wood is funeral pyre of rich people, a taboo as fuel. Tree is symbolically sacred, wish granting tree and tree motif of Shiva Temple. Leaves are offered to Lord Shiva. Fruits are believed as head of Shiva.

(Jain, 1991)

Critical Notes

Uses of fruits for throat complaints and cough respectively are still to be reported.

***Aerides crispum* Lindl.**

Family: Orchidaceae, Local Name: Vinchu, Bandgool

Ethnobotanical Notes

Medicine

1. Seed powder is applied on injuries of human body. It cures injury. It is collected preferably from mango tree, being epiphytic.

Critical Notes

The taxon is being reported for the first time as useful. Seeds are medicinally beneficial to treat injuries.

****Aerva lanata* (L.) A.L.Juss. ex Schult.**

Family: Amaranthaceae

Ethnobotanical Notes

Miscellaneous

1. Leaves are used as fodder for goats and cattle.

Notes from Classics

Roots are useful on head-ache. Spikes are used for stuffing pillows.

(Watt, 1889–1893)

Plant are diuretic and demulcent.

(Anonymous, 1948–1976)

Herb is diuretic and demulcent. Plant is given to herds and camels as a purgative and emetic.

(Ambasta, 1986)

Roots are used on boils. Leaves are used on ear complaints and also eaten as vegetable. Plant is used on cholera, diarrhoea, dysentery and white discharge during urination. Plant is used as and antidote on snake-bite.

(Jain, 1991)

Critical Notes

Consumption of leaves by pet animals is a new notice by the present investigators.

Aeschynomene indica L.

Family: Papilionaceae, Local Name: Junglichinch

Ethnobotanical Notes

Miscellaneous

1. The shoots are used as fodder for goats.

Notes from Classics

Pith of plant is used as substitute for corks in medicine bottles.

(Watt, 1889–1893)

Plant is used as fodder but toxic for horses in fruiting stages. Seeds yield fatty oils

(Anonymous, 1948–1976)

Entire plant is used in fever and impotency. Wood charcoal is used as gun powder.

(Jain, 1991)

**Agave cantala* Roxb.

Family: Agavaceae, Local Name: Chopadi Ghaypat

Ethnobotanical Notes

Veterinary Medicine

1. Leaf powder is mixed with wheat flour and fed to cows and buffaloes to increase lactation.

Notes from Classics

Leaf fibres are used for rope making. It is planted for hedge around field. Leaves juice and flower juice of *Eluesine corcana* Gaertn. is applied to contusions of drought–cattle.

(Watt, 1889–1893)

Leaves yield fibre used for rope, mats, weavers, cordage, twines, nets etc.

(Anonymous, 1948–1976)

Leaves yield fibres which is used in hard fibre twines.

(Ambasta, 1986)

Roots are used on boils, fever, goitre, paralysis and skin diseases. Leaf paste is applied on burns, cuts, and wounds, also used to control diarrhoea. Leaves yield fibre. Leaves externally useful on skin diseases of cattle.

(Jain, 1991)

Critical Notes

Leaf powder to increase lactation in case of pet animal is a new application. It is exotic, native of America. It is introduced especially for fibres.

Agave vera-cruz Mill.

Family: Agavaceae, Local Name: Ghaypat

Ethnobotanical Notes

Miscellaneous

1. Inflorescence axis is cut longitudinally and used for making shutters of huts.
2. Plants are planted as a hedge on the boundries of fields.

Notes from Classics

Fibre is pulled from leaves, used for making ropes, mats and cordage.

(Anonymous, 1948–1976)

Leaves yield fibres which are used for rope making. Soft wood portion of stem is cooked with Tamarind fruit and jaggery and are consumed by poor.

(Ambasta, 1986)

Plant yields fibres, used for cordage, making baskets, brooms, fish-nets, and stuffing and paper manufacture.

(Jain, 1991)

Critical Notes

The plant species is usually planted for fibres and for prohibiting animals in the cropped fields. The tribal people have also made appropriate use of the gigantic inflorescence axis which is otherwise wasted. It is exotic and denizen of American deserts.

**Ageratum conyzoides* Linn.

Family: Compositae, Local Name: Boradu

Ethnobotanical Notes

Medicine

1. Two to three leaves are boiled in water. This decoction is given on a cramp in abdominal pain.

Notes from Classics

Leaves are used on cuts and sores as external application in ague (Fever).

(Anonymous, 1948–1976)

Leaves are styptic and vulnerary, also yield essential oil. Extract of flowers in used in leukaemia. Seeds yield fatty oil. Juice of entire plant is useful in prolapsus. Decoction or infusion of herb is used in diarrhoea, dysentery, colic with flatulence and other gastro-intestinal ailments.

(Ambasta, 1986)

Roots are used in the treatment of kidney stones. Leaves are antiseptic, used on boils, burns, cancer, cuts, leprosy and skin diseases. Entire plant is used in diarrhoea, febrifuge, colics, headache, muscular pains, piles, ring-worm, scabies, sores, tumour, and uterine disorders. Plant is used against snake poison.

(Jain, 1991)

Critical Notes

Application of leaves in abdominal cramp is new. The plant species in exotic, native of South Africa.

Ailanthus excelsa Roxb.

Family: Simaroubaceae, Local Name: Maharukhi

Ethnobotanical Notes

Medicine

1. Leaves are crushed and tablets are prepared from paste. One tablet along with banana ripe fruit is given for seven days two to three times a day to control jaundice.

Veterinary Medicine

1. Leaf juice is given for five days for animal flatulence. It is given in large quantities.

Notes from Classics

Stem bark is useful aromatic and dyspeptic complaints, as a tonic and febrifuge in cases of debility, as expectorant and antispasmotic, also given in chronic bronchitis and asthma. In Bombay, bark and leaves are in great repute as a tonic especially in debility after delivery. Powder of resins in used to control dysentery and cough, also used as astringent in dysentery and diarrhoea.

(Watt, 1889–1893)

Stem bark is used as febrifuge and tonic. Timber is used for packing cases, fishing floats, boats, spear sheaths, sword handles, toys and drums.

(Anonymous, 1948–1976)

Stem bark is anthelmintic, febrifuge, expectorant, and antispasmodic, used for asthma, bronchitis and also in dysentery. Leaves and shoots are used as fodder. Wood is used for catamarans, and small boats also used for veneers, plywood packing cases, sword handles, toys and poor quality matches.

(Ambasta, 1986)

Stem bark is useful in cold, cough, bronchitis, febrifuge, infertility and rheumatism. Leaves are useful in skin diseases. Stem bark and leaves are used as tonic. Stem bark is used for wounds and swelling of cattle.

(Jain, 1991)

Critical Notes

Medicinal uses of leaves against jaundice and their veterinary application to treat flatulence are new and noteworthy.

Albizia lebbeck Benth.

Family: Mimosaceae, Local Name: Shiras

Ethnobotanical Notes

Miscellaneous

1. Leaves are used as cattle fodder.
2. Wood is used for house construction, doors and windows.

Notes from Classics

Root bark powder is used for ulcerated gums and also to strengthen them. 'Anjan' prepared from seeds is used for ophthalmic diseases. Seeds are astringent and given in cases of piles, diarrhoea, and gonorrhoea. Seed oil is considered to be useful in leprosy. Seed powder or seed paste is administered in cases of scrofulous enlargement of glands. Leaves are used as camel fodder. Wood is used for picture frames, sugar mill, oil-mill crushers, furniture, boards and well curbs. In northern India, timber is used for house construction.

(Watt, 1889–1893)

Leaves are lopped for fodder and green manure. Wood is used for furniture, constructions of houses, railway carriages, agricultural implements, oil and cane pressers, carts, well curbs, and carving etc. Wood is used as fuel for cooking purposes.

(Anonymous, 1948–1976)

Stem bark is used on boils. Leaves and seeds are used on eye complaints. Stem bark is used for tanning fishing nets. Wood is excellent for furniture, internal decoration, house constructions, railway carriages, agricultural implements, oil and cane crushers, pressers, well curbs, carts, and carvings. Gum is used to adulterate gum of *A.arabica* (*A.nilotica*)

(Ambasta, 1986)

Roots are useful on gum ulcers. Leaves are used on night blindness. Stem bark is used on eye sores. Flowers are useful in carbuncle, boils and swellings. Seed are helpful for piles. Seeds and bark are used in diarrhoea, dysentery and gonorrhoea. Symbolically God of Love 'Kāmadeva' uses flowers on arrows of desire.

(Jain, 1991)

**Alocasia macrorrhiza* Schott.

Family: Araceae, Local Name: Brahma- rakshas

Ethnobotanical Notes

Medicine

1. Leaf juice is applied for 4–5 days on foot to treat swellings.

Notes from Classics

Stem juice is reported to be used to get relieve on scorpion sting as well as nettle sting. Roots and leaves are used as rubefacient for joint pains

(Anonymous, 1948–1976)

Roots are considered to be laxative and diuretic. Aerial stems along with tuberous underground portions and petioles are eaten as vegetable. Acrid juice is applied to get relief from sting of giant nettle.

(Ambasta, 1986)

Corm is edible as vegetable.

(Jain, 1991)

Critical Notes

Leaf juice is being reported for the first time for treating swelling of foot.

**Aloe vera* (L.) Burm. *f.*

Family: Liliaceae, Local Name: Korpad

Ethnobotanical Notes

Medicine

1. Leaf sap and fruits are rubbed in sap and the slurry obtained is applied on injury of dog bite.

2. Leaf powder is applied on burns.
3. Leaf pulp is kept on eyes to get relief from burning sensation.

Notes from Classics

Root is supposed efficacious in colic. Fresh leaf juice is said cathartic, cooling and used in fevers, spleen and liver diseases, enlarged lymphatic gland, also externally applied on eye lids for eye diseases. Leaf pulp is applied on boils. Leaf juice acts powerfully on uterus, and useful as an emmenogogue.

(Watt, 1889–1893)

The plant is a valuable treatment in piles and recital tissues. Mucilage is cooling and used to poultice inflammations.

(Anonymous, 1948–1976)

Fresh juice of leaves is cathartic and refrigerant, used in liver and spleen ailments and for eye troubles, also useful in X-ray burns, dermatitis, cutaneous leishmaniasis and other skin diseases. Entire plant yields dye and fibre.

(Ambasta, 1986)

Leaves are used on boils, hardening of breast tissue, constipation, jaundice, fever, gonorrhoea, menstrual complaint, liver complaints, piles, rheumatism, sexual vitality, skin diseases and spleen complaints. Leaf juice is used as hair wash. Leaves are used as veterinary medicine.

(Jain, 1991)

Critical Notes

Leaf sap is not reported beneficial in case of dog-bite in past. It is exotic, native of North America.

**Alstonia scholaris* R.Br.

Family: Apocynaceae, Local Name: Saitan, Satvin

Ethnobotanical Notes

Medicine

1. Stem bark is burnt and a blanket (Ghongadi) is passed

over the smoke and the same is wrapped over body of a person suffering from itching (Khujali).

Notes from Classics

The stem bark is used as astringent tonic, anthelmentic, alteratives and antiperiodic. It is a valuable remedy in chronic diarrhoea and advanced stages of dysentery. It is also useful in catarrhal fever. Milky latex is applied on ulcer. Milky latex mixed with coconut oil is poured into ears in case of ear-ache It also restores tone of stomach in debility or after fever. Tender leaves are roasted, pulverized and made into poultice used to unhealthy ulcers with foul discharge. Wood is used for boxes, furniture, scabbard, coffins, in Burma used for blackboards.

(Watt, 1889–1893)

Latex is applied on sores and ulcers. Wood is used for packing cases, tea-boxes and minor furniture, and in plywood match box manufacturing.

(Anonymous, 1948–1976)

Stem bark is bitter tonic, febrifuge, anthelmentic and galactogogue, used in the form of liquid extract or tincture for chronic diarrhoea, asthma, and cardiac troubles, also used as an haemostatic. Stem bark yields fibre. Leaves are used in beri-beri, dropsy, and in congested liver. Latex is applied on sores, ulcers, tumours, and rheumatic swellings. Wood is used for packing cases, teaboxes, writing boards, furniture, plywood, match splints, and paper industry.

(Ambasta, 1986)

Stem bark is used in case of chest pain, diarrhoea, eye complaint, fever, malaria, pneumonia and rheumatism entire plant is used in asthma, bronchitis, rheumatism, head-ache, nueralgia, pain in leg and also used in cholera. Latex in useful on tuberculosis and ulcers in mouth. Leaves are used on gastric diseases. Stem bark is used as antidote on snake bite. Miscellaneous ornaments of twigs are worn as rings during folk dances.

(Jain, 1991)

Critical Notes

Smoke of stem bark to treat itching is a new application.

Alternanthera sessilis DC.

Family: Amaranthaceae, Locale Name: Tandalya, Jalgambha

Ethnobotanical Notes

Miscellaneous

1. Leaves are cooked and eaten as vegetable.
2. Plant is used for fodder for cows, bullocks and buffaloes.

Notes from Classics

Leaves are eaten as vegetable by mothers to increase flow of milk, also used as a wash for the eyes.

(Watt, 1889–1893)

Plant is used as pot-herb.

(Anonymous, 1948–1976)

Plant is used as cattle fodder, increases the flow milk in cattle, and also used for night blindness.

(Ambasta, 1986)

Spray of root juice repels bees. Leaves are applied on bone-fracture, dysentery, malarial fever, night blindness, post-natal and puerperal fever. Plant is used as an veterinary medicine. Plant is used as antidote on snake-bite, rabid dog-bite or jackal bite.

(Jain, 1991)

***Alternanthera tenella* Colla

Family: Amaranthaceae, Local Name: Korda-ganja

Ethnobotanical Notes

Medicine

1. Inflorescence is burnt and the ash, called 'Misri' is wrapped in the leaf of *Piliostigma malabaricum* (Roxb.) Benth. This bidi is smoked. Smoke is passed through an ear as a treatment for ear-ache.

Critical Notes

It is exotic, denizen of Tropical America. The plant has recently naturalized in this region. Still the tribals have tried it for medicinal purpose as given above. This is an additional use.

Alysicarpus rugosus (Willd.) DC.

Family: Papilionaceae, Local Name: Shevra

Ethnobotanical Notes

Medicine

1. Entire plant is used as fodder for domestic animals.

Notes from Classics

Entire plant is used as goat fodder.

(Anonymous, 1948–1976)

Plant is good fodder for goats, green and dry.

(Ambasta, 1986)

Seeds are cooked and eaten.

(Jain, 1991)

Amaranthus spinosus L.

Family: Amaranthaceae, Local Name: Kate-math

Ethnobotanical Notes

Medicine

1. Leaves of this plant and bark of *Artocarpous heterophyllus* Lam. are burnt together. This ash is mixed with coconut oil. It is then applied to cure eczema.

Miscellaneous

1. Leaves are cooked and eaten as vegetable.
2. Entire plant is fed to cattle.

Notes from Classics

Root is an excellent remedy for gonorrhoea. Roots made into poultice are applied to baboes and abscesses for hastening suppuration. Poultice of leaves is considered as lactogogue and boiled with pulse is given to cows. Leaves are employed for colic pain on scorpion bite. Entire plant is used as antidote for snake-poison. Plant is used as pot-herb.

(Watt, 1889–1893)

Plant boiled with pulse is used as pot-herb. Plant is fed to cattle to increase lactation.

(Anonymous, 1948–1976)

Plant is used as pot-herb, considered as sudorific, febrifuge and recommended for eruptive fevers. Leaves are emollient. Infusion of shoots is useful for eczema. Plant is also used as lactogogue.

(Ambasta, 1986)

Roots are abortifacient, laxative, used in colics, eczema and gonorrhoea. Seeds are useful on cold and cough, also reduce body heat. Roots are used as anti-dote against snake-bite and scorpion sting. Entire plant is anthelmintic, helpful in blood purification, piles and is said to be purgative.

(Jain, 1991)

Amaranthus viridis L.

Family: Amaranthaceae, Local Name: Matla

Ethnobotanical Notes

Miscellaneous

1. Leaves are cooked as vegetable.
2. Entire plants constitute a good fodder for pet animals.

Notes from Classics

Tender tops are used as vegetable by natives.

(Watt, 1889–1893)

Plant is used as pot-herb, and eaten as spinach. Plant is used as cattle fodder.

(Ambasta, 1986)

Leaves are eaten to improve digestion. Leaves are vermifuge and eaten as vegetable. Leaf juice is used on scorpion or centipede bite.

(Jain, 1991)

**Amorphophallus commutatus* (Schott.) Engl.

Family: Araceae, Local Name: Shewalkand

Ethnobotanical Notes

Medicine

1. Fruits are burnt and this ash, called 'Misri' is homogenized with coconut oil. It is then applied on piles till cure.

Miscellaneous

1. Tubers are cooked and consumed as vegetable.

Notes from Classics

Tubers, leaves and flowers are used as vegetable. Tubers are useful against snake-bite.

(Jain, 1991)

Critical Notes

Ash of fruits is useful in treating piles. This is unreported yet.

**Ampelocissus latifolia* Planch.

Family: Vitaceae, Local Name: Bendore

Ethnobotanical Notes

Medicine

1. Roots are chewed or one tea spoon juice of root is given for easy delivery.
2. Stem is burnt and the ash called 'Misri' is applied on stomach of a pregnant women to get relief from stomach pains.

Miscellaneous

1. Stems are splitted into fibres and used for tying grass or wood bundle.
2. Caution: Goats die if the plant is consumed by them.

Notes from Classics

Decoction of roots is given in dysentery. Leaf juice is used on dental troubles and also used as detergent on indolent ulcers.

(Ambasta, 1986)

Roots are used on boils, to treat bone fracture, dysentery, menstrual complaints, stomach pains, pneumonia, skin diseases, wounds and as galactogogue. Root bark is used in fever. Fruits are edible. Roots are used on snake-bite.

(Jain, 1991)

Critical Notes

Roots for easy delivery, stem for stomach pains and to tie fodder of fuel wood bundles are new reports. The observations of tribals about consumption by goats and its bad effect is notable.

**Anacardium occidentale* L.

Family: Anacardiaceae, Local Name: Kajuamba

Ethnobotanical Notes

Medicine

1. Seeds are rubbed on stone and this slurry is applied on injury caused by scorpion sting.

Miscellaneous

1. Riped thalamus is like ribals.

Notes from Classics

Seed oil is used as anesthetic in leprosy and on blisters, warts, corn and ulcer. Fruits juice is used as stimulant after distillation. Fruits are eaten raw as a treatment on scurvy. Kernels are fried and eaten, also made into confectionary with sugar. Stem juice is used for marking ink. It is astringent, used by natives as a flux for soldering metal. Ripe fleshy stalks called 'torus' are eaten as fruits. Bark juice is used for tanning. Juice of nuts is used as a substitute for iodine. Kernel oil is nutritious and superior to olive oil. Oil obtained from pericarp or shell of nuts is used to colour and preserve fishing lines in Andaman. Wood is used for packing cases.

(Watt, 1889–1893)

Fruits are edible. Cashew apple juice can be converted into vinegar. Peels of kernels are utilized as poultry food. Gum is used for book binding. Sap of bark is used for indelible ink for marketing lenin.

(Anonymous, 1948–1976)

Cashew apple is pseudo-fruit, used for jams, and preparations of wines and beverages. Cashew kernels are eaten raw or fried, used as a dessert and employed in sweat-meats and confectionary. Kernels are highly nutritious. Cashew shells yield a drying oil, used for water proofing, preservative in painting of boats, fish-nets and light wood work. Cashew oil is an excellent lubricant in magnet armatures in aeroplanes, also used in varnishes, inks, termite proofing timber and insulation coatings. Gum is used is varnishes and protection of books and light wood work against insects. Milky sap is used as indelible ink for marking. Wood is used for packing cases and boats.

(Ambasta, 1986)

Roots are purgative. Stem bark is used in diarrhoea, leprosy, skin diseases, kernels are edible and useful as tonic.

(Jain, 1991)

Critical Notes

Seeds are found useful on scorpion sting. This is not reported in past classical literature. It is exotic tree species, native of Tropical America. It is introduced for its valuable fruits.

**Anogeissus latifolia* (Roxb. ex DC.) Wall. ex Bedd.

Family: Combretaceae, Local Name: Dhamoda

Ethnobotanical Notes

Medicine

1. Stem bark powder is mixed thoroughly in water. This filtrate is given against dysentery.

2. Paste of stem bark is applied on wounds.
3. Infusion of stem bark is given on chronic cough (Dangya khokla). Half cup is given daily till cure.
4. Stem bark is scrapped by teeth and crushed into paste which is applied on injury of scorpion sting.
5. Stem bark of this plant and Sawar (*Bombax ceiba* Linn.) are crushed and kept in water. Infusion of 20–25 ml twice a day is useful against dysentery. It is advised till cure.
6. Powder of stem bark and its gum wrapped in the leaf of *Piliostigma malabaricum* (Roxb.) Bth. and this bidi is smoked to get relief from this scorpion bite.
7. Gum is consumed by tribals.

Notes from Classics

Gum is used in the treatment of cholera. Leaves yield black dye, used for tanning in Bombay. Gum is used in calico printing. Wood is used for cart making, furniture, sleepers, ship building, agricultural implements, and as valuable fuel. It is an excellent charcoal yielder.

(Watt, 1889–1893)

'Tasar' silkworms are sometimes fed on its leaves. Tender leaves are used for tanning in small leather industry. Gum is substitute to gum of [*Acacia nilotica* (L.) ssp. *indica* (Bth.) Brenan]. Wood is used for agricultural implements, furniture and building etc.

(Anonymous, 1948–1976)

Ambasta (1986) mentioned uses similar to those in Anonymous (*loc.cit.*).

Stem bark is used on colics, cough, diarrhoea, dysentery, headache, liver complaints, skin diseases, also used as antidote on scorpion bite. Stem bark and roots are used against stomach complaints. Gum is edible and administered as tonic. Tree is used in the treatment of cancer, on face, carbuncle, cholera, cold, colic, cough, dysuria, excess perspiration, spleen complaint, and in toothache.

(Jain, 1991)

Critical Notes

The uses of stem bark against dysentery, on wounds and 'Dangya Khokla' are additional.

**Annona squamosa* L.

Family: Annonaceae, Local Name: Sitaphal

Ethnobotanical Notes

Veterinary Medicine

1. Leaf paste is used to cure injuries, and septics. It is also used to repel worms from wounds of cattle.

Miscellaneous

1. Ripe fruits are edible.

Notes from Classics

Root is employed in depression of spirits and spinal diseases. Stem bark is employed in asthma and fever. Infusion of leaves is efficacious in prolapsus ani of children. Leaves are applied for extraction of guinea worms. Paste of green leaves without adding water is applied on unhealthy ulcers. Leaves are used for anthelmintic. Seeds when applied to the Osuteri causes abortion. Pounded leaves alongwith tobacoo leaves and quicklime are used in ill conditioned ulcers to destroy maggots in animals, also applied to sores infested with maggots. Ripe fruits when bruised and mixed with salt are applied to malignat tumours to hasten suppurations. Dried unripe fruit powder is mixed with gram flour, and used to destroy vermin. Seeds show poisonous action on lice.

(Watt, 1889–1893)

Roots are used for destroying insect and lice. Leaves are used as poultice to produce suppuration. Seeds are abortificient.

(Anonymous, 1948–1976)

Roots are purgative. Seeds are abortificient. Fruits are edible, various drinks and liquor can be prepared from fruits.

(Ambasta, 1986)

Roots are used on melancholia and spinal disorders. Leaves are used against guinea worms and lice. Leaves and seed powder is used in hysteria or leaves are boiled and vapours are inhaled. Fruits are used on dysentery, also used to treat tumour. Seeds are used in menstrual complaints. Leaf paste is applied on wounds, ulcers and worms in septics.

(Jain, 1991)

Critical Notes

Leaves for curing injuries, to avoid septic and to repel worms from wounds in case of pet animals are new applications of this taxon. It is exotic, native of Tropical America. It has also naturalized at some places in this region.

Apluda mutica Linn.

Family: Graminae, Local Name: Holera, Phulia

Ethnobotanical Notes

Miscellaneous

1. Entire plants are used for thatching roofs of huts.
2. Plants are used as fodder for cows, bullocks and buffaloes.

Notes from Classics

Stem stalks are used for making hats. Grass is used as raw material for rayon industry. Young grass is liked by buffaloes.

(Watt, 1889–1893)

Entire plant is used as medicine for mouth sores of cattle.

(Jain, 1991)

Argemone mexicana L.

Family: Papaveraceae, Local Name: Kardai

Ethnobotanical Notes

Miscellaneous

1. Seed oil is used for cooking purposes by native Bhils.
2. Yellow latex is used on skin diseases.

Notes from Classics

Root powder is useful in treating tape-worms. Seeds have narcotic properties. Seed oil is applied externally for head-ache and skin diseases. Yellow latex of plant is used for dropsy, jaundice and cutaneous affections. It is diuretic and removes blisters, heats and ulcers. Juice is used on indolent ulcers, and removes specks on cornea. Juice mixed with ghee is given in gonorrhoea, itch and scabies. Oil is used on emulsyfier for lamps. Oil along with sugar relieves the stomach pain and gives refreshing sleep. Oil is valuable aperient in dysentery.

(Watt, 1889–1893)

Seed oil is useful remedy for skin diseases. Latex is used externally in scabies and opthalmia. Seed oil mixed with dyeing oil and linseed oil is used in paint industry.

(Anonymous, 1948–1976)

Seeds yield nauseous, bitter, non-edible oil, useful in cuteneous troubles, it is cathartic. Presence of argemone oil in edible mustard oil is responsible for out breaks of epedemic dropsy. Seed oil is used in paint industry and soap making. Yellow latex is applied for scabies and also helps in opthalmia.

(Ambasta, 1986)

Roots are used on piles, inflammation and as mouth wash. Leaves are used for boils, ulcers, syphilis, leprosy and dropsy. Seeds are used for asthma, cough, whooping cough, head-ache, caries and gum troubles. Latex is used against jaundice, skin diseases, scabies, tooth-ache and eye complaints. Roots are useful against dog-bite. Stem pieces are cooked as vegetable.

(Jain, 1991)

Critical Notes

Report of seed oil consumption is unhealthy for human being. Its adulteration in mustard oil in Uttar Pradesh and Bihar made havoc few years ago. It is also documented by Ambasta (*loc.cit.*). However, it is used by the Bhils as such but no bad effect is reported by them. It is exotic weed, native of Mexico.

***Argyeria sericea* Dalz.

Family: Convolvulaceae, Local Name: Ekota

Ethnobotanical Notes

Veterinary Medicine

1. Leaf paste is applied on neck or shoulder of bullocks to cure swellings caused due to yoke.

Critical Notes

The taxon is being communicated as useful one for the first time.

**Arisaema murrayi* (Grah.) Hook.

Family: Araceae, Local Name: Nagphani

Ethnobotanical Notes

Miscellaneous

1. Tubers are boiled thoroughly and eaten as food especially in times of scarcity.
2. Inflorescences are eaten as vegetable. Only the core of inflorescence axis is cooked.

Notes from Classics

Tubers, flowers, and fruits are eaten as vegetable.

(Jain, 1991)

Critical Notes

The use of core of inflorescence is an additional application as vegetable.

***Arthraxon lancifolius* (Trin.) Hochst.

Family: Graminae, Local Name: Turda

Ethnobotanical Notes

Miscellaneous

1. Fodder: Plant is used as cattle fodder.

Critical Notes

This grass species is not noted earlier as a source of fodder.

**Artocarpus heterophyllus* Lam.

Family: Moraceae, Local Name: Phanas

Ethnobotanical Notes

Medicine

1. Stem bark of this plant and leaves of *Amaranthus spinosus* L. are burnt together. This ash is mixed with coconut oil. The paste is applied on skin for curing eczema.
2. Unripe fruits naturally dropped from tree are dried and powdered. One spoonful of it is taken orally two control vomiting.
3. Prickles of fruits are removed and burnt. This 'misri' (ash) is applied on tongue or throat infection or sores. It is advised once daily till cure.
4. Seed pulp of ripe fruits is edible.

Notes from Classics

Roots are internally useful in diarrhoea. Young roots are used on skin diseases. Unripe fruits are astringent. Ripe fruits are laxative. Fruits on distillation yield an alcoholic beverage. Juice of entire plant is externally applied on glandular swellings and abscesses to promote suppuration. Juice from bark is used as valuable bird-lime and as cement. Wood or saw dust yield a yellow dye to colour to Burmese priests robes. Unripe fruits are cooked in curry with shrimps. Seeds are roasted and eaten. Seed flour is made into cakes and eaten. Stem bark yields fibre. Wood is used for making furniture.

(Watt, 1889–1893)

Unriped fruits are cooked and eaten as vegetable or also pickled. Ripe fruits are eaten fresh or preserved into syrups. Wood is used for general carpentry works, furniture and musical instruments. Extract of heart wood with alum is used by Buddhist priests for dyeing their robes yellow.

(Anonymous, 1948–1976)

Ambasta (1986) reported uses similar to those in Anonymous (*loc.cit.*).

Roots are used on stomach complaints. Latex is applied for skin diseases. Latex is used to seal holes on earthen pots. Whole plant is used for carbuncle, post-natal, small pox, sores, sterility, stomach complaints and ear-ache. Fruits are edible. Yellow dye in extracted from wood. Wood is used for furniture and musical instruments.

(Jain, 1991)

Critical Notes

The use of stem bark on eczema and of fruits to control vomiting and for tongue or throat sores or infection are yet to be reported.

**Asparagus racemosus* Willd.

Family: Liliaceae, Local Name: Sashachi ghongadi

Ethnobotanical Notes

Miscellaneous

1. Tuberous roots are fed along groundnut cake to milching cows and buffaloes to increase flow of milk.
2. Roots are boiled and consumed.

Notes from Classics

Roots are used as refrigerant, demulcent diuretic, aphrodisiac, antispasmodiac, alterative, antidiuretic, and antidysentery. It prevents confluence of small-pox. It is used in the form of preserve in impotence. Roots are chiefly used as demulcent in veterinary.

(Watt, 1889–1893)

Honey is mixed with root juice and given as a demulcent in dyspepsia. Medicated oil is prepared from roots and useful for nervous and rheumatic complaints.

(Anonymous, 1948–1976)

Root juice is mixed with honey and given in dyspepsia. Root oil is used in nervous and rheumatic complaints. Herb is used as a tonic, diuretic and galactogogue.

(Ambasta, 1986)

Tuberous roots are used in the treatment of rheumatism, bleeding nose, blood in urine, dysentery, diarrhoea, gastric complaints, ulcers on tongue, wounds, menstrual complaints, gonnorhoea and leucorrhoea. Fruits are used against stomach-ache. Plant is useful for impotency, and in urine complaint. Plant used as antidote for snake-bite. Tubers are used as veterinary medicine. Entire plants are symbolically used in God's shrine at the time of Hareli festival. Plants keep off ghosts. It is also used in religious ceremonies of Santals.

(Jain, 1991)

Critical Notes

Tuberous roots to increase flow of milk in case of milching animals is a new application of the plant.

**Aspidopterys cordata* (Heyne ex Wall.) A.Juss.

Family: Malpighiaceae, Local Name: Shrivel, Makadvel

Ethnobotanical Notes

Medicine

1. Roots are chewed to treat stomach-ache.
2. Latex is applied to rice grains and fed to mothers for one to two days to increase lactation.

Miscellaneous

1. Fodder: Leaves are fed to cows and buffaloes to increase lactation.

Critical Notes

All above mentioned uses have remained unnoticed in past.

**Azadirachta indica* A.Juss.

Family: Meliaceae, Local Name: Kadu Limb

Ethnobotanical Notes

Medicine

1. Dried stem bark is burnt and mixed with coconut oil is applied on wounds.
2. Leaf juice is dropped in ears for ear-ache.

3. Decoction of leaves is added in bath water. It is used for bath by person suffering from fever.
4. Leaves are given to person of snake bite, creates vomiting and reduces poison.

Notes from Classics

Root bark and young fruits are tonic and antiperiodic. Root is bark is useful in intermittent fever and debility Bark is useful in fever, thirst, nausea, vomiting and skin diseases. Leaf ash mixed with ghee is externally applied to psoriasis. Fresh juice of leaf is given with salt in cases of intestinal worms and with honey in skin diseases and jaundice. Leaf pulp is applied to breasts, arrest the secretion of milk. Poultice of leaves is used on ulcerations and skin diseases. A hot infusion of leaves is used for formenting glands swollen glands, bruises, sprains and appears to be anodyne. Flowers are stimulant and tonic. Flowers are used in atonic dyspepsia and general debility. The oil, seeds, and leaves are local stimulant, insecticide and antiseptic. Gum is demulcent tonic. Seed oil is used for destruction of lice and as application in urticaria and eczema. The toddy is refrigerant, nutrient and alterative-tonic. Leaves are cooked as vegetable in form of curry or are simply parched and eaten. Wood is used for house constructions, carts and making agricultural implements.

(Watt, 1889–1893)

Stem bark is used on skin diseases. Leaves are boiled and applied in form of poultice for skin affections. Decoction of leaves is recommended in ulcers and eczema. Fresh tender twigs are used for brushing teeth particularly in case of pyorhoea. Shade dried leaves are kept in clothes and books, to protect from paper moths etc. Neem oil is used in manufacture of soap. Dried flowers are considered tonic and stomachic. Seed oil is used on skin diseases, indolent ulcer, sores, ringworms, scrofula and rheumatism. Oil cake is regarded as useful fertilizer. Timber is used for house construction, ploughs etc.

(Anonymous, 1948–1976)

Uses reported by Ambasta (1986) are similar to those in Anonymous (*loc.cit.*).

Stem bark is used in antifertility in males, also used on boils, jaundice, liver complaints, malaria, wounds and skin diseases etc. Leaves are antiseptic, used on diabetes, diarrhoea, dysentery, dyspepsia, gout, heart complaints, scabies and skin diseases. Leaves are burnt as insect repellant. Leaves are chewed in cases of snake-bite. Flowers are used in dyspepsia and skin diseases. Seeds are used as blood purifier and post-natal. Fruits are used on leprosy, piles, skin diseases, urine complaints and as vermicide. Entire plant is useful in the treatment of cancer, cholera, cough, cold, dysentery, gum inflammation, measles, rheumatism, small-pox, spleen complaints, syphilis, tumour and ulcers. Buds, flowers and fruits are edible. Tree is symbolically sacred. Shrines are made under it. These are planted near temples. Idols of god are made from woods, Its deity in sheetlamata, Goddess of small-pox, keeps evil spirits away, in snake cult. Tree is spirited for muslim saints, they rest under it.

(Jain, 1991)

Critical Notes

Leaves are being reported useful for the first time for ear-ache and fever. Plant is exotic, native of Persia..

*_Bacopa monnieri_ (L.) Penn.

Family: Scrophulariaceae, Local Name: Bam

Ethnobotanical Notes

Medicine

1. A glass of plant extract is given orally at night after meals to reduce excess body heat.
2. Leaf juice, 2 to 3 spoons per day is given in fever till cure.

Notes from Classics

Powder of dried leaves is given in cases of asthenia, nervous break-down, and directly used as cardio tonic. Entire plant is used in indigenous medicine as a nerve tonic and also to cure epilepsy and insanity. Medicated ghee is prepared from plant, and used in hysteria.

(Anonymous, 1948–1976)

Leaves are used as diuretic and aperient. Plant is said to improve intellect, used on epilepsy, insanity and other nervous complaints.

(Ambasta, 1986)

Leaves are useful in the treatment of abdominal pains, cough, rheumatism, and also to treat lives complaints. Plant is useful in blood purification, boils, epilepsy, head-ache, insanity, stomach complaints, ring worms, and used as nervine tonic, general tonic and hair tonic. Plant juice is useful in fever. Plant extract is orally administered for seven days to reduce excess heat from body or used as febrifuge.

(Jain, 1991)

Critical Notes

Plant extract to reduce body heat and leaf juice to treat fever are additional reports of application of this species.

Balanites aegyptiaca (L.) Delile

Family: Balanitaceae, Local Name: Hingan-bet.

Ethnobotanical Notes

Miscellaneous

1. A stem of a plant growing near ant-hill is collected at morning especially on Saturday. It is cut into small pieces. Bundles of these pieces are kept in a house or shop for better future. The bundles are tied on the legs of domestic animals for healthyness.

Notes from Classics

Stem bark is anthelmintic. Seeds, fruits, bark and leaves are used as a native medicine. Bark is useful as veterinary medicine. Bark juice is used as fish poison. Fruits are eaten ripe. Fermented fruit yield toxicating liquor used by Negroes. Nuts are employed in fire works. They are used in crackers. Fruit pulp is used as detergent to clean silk in Rajputana.

(Watt, 1889–1893)

Stem bark is anthelmintic for cattle. Bark juice is used as fish poison. Fruit pulp is edible. Fruits are used for washing silk clothes.

Nuts are made into crackers by drilling small holes. Kernels or seed oil is used for soap making in Uganda. Stems are used as walking sticks.

(Anonymous, 1948–1976)

Stem bark is anthelmintic. Unripe fruits are cathrtic, ripe fruits are used for whooping cough on skin complaints. Seeds are expectorant, used on cough and colics. Seed oil is applied on burns, freckles, and also used in soap making. Wood is used as walking sticks.

(Ambasta, 1986)

Stem bark is used on swollen face and urine complaint. Bark and fruits are used as vermifuge. Fruits are purgative, and used on pneumonia and skin diseases. Seed powder is applied on burns. Seed oil is useful for cough. Fruits and roots are used as antidote on snake-bite. Fruit juice is used as detergent. Fruits shell are used in fire-works.

(Jain, 1991)

Critical Notes

The plant is already reported useful as veterinary medicine. Keeping the stem pieces in abodes or shops for better future of human being is an ethnobelief.

**Baliospermum montanum* (Willd.) Muell.-Arg.

Family: Euphorbiaceae, Local Name: Sapedi

Ethnobotanical Notes

Medicine

1. Roots are crushed and two to three spoons of extract is administered as a remedy to avoid vomiting.
2. Stem pieces are used as tooth brush.

Notes from Classics

Roots are considered cathartic and are administered with aromatics in dropsy, anasarca and jaundice. Leaves are used to treat wounds. Their decoction is useful in asthma. Seeds are drastic purgative, but over doses are acronarcotic poison. It is also used as a substitute for *Croton tigulium* (Jamalgota). Seed oil is externally used on rheumatism.

(Watt, 1889–1893)

Roots are cathartic. Seed oil is powerful hydrogogue, cathartic, externally used on rheumatism.

(Anonymous, 1948–1976)

Ambasta (1986) reported similar uses as those in Anonymous (*loc.cit.*).

Roots are laxative used in the treatment of dropsy and jaundice. Stem is used on tooth-ache. Leaves are employed in case of asthma. Seed oil is used on rheumatism, gout, and gastric complaints. Seeds are used as an antidote- on snake bite. Roots are used as veterinary medicine.

(Jain, 1991)

Critical Notes

Use of roots to avoid vomiting and stems as tooth brush are not recorded earlier.

Bambusa arundinacea (Retz.) Willd.

Family: Graminae, Local Name: Kaset, Tokar

Ethnobotanical Notes

Miscellaneous

1. Leaves are used as fodder.
2. Stems are used for roofing huts.
3. Strips of stems are employed for making fish and crab traps and also used for making various articles.
4. Fibre is extracted from culms.

Notes from Classics

Poultice of young shoots are applied for the dislodgement of worms in ulcers. Juice is first poured on the vermin and the ligneous mass is applied and secured by a bandage. Decoction of leaf-bud is used to encourage the free discharge of menses or lochia, also used in leprosy, fevers and hoemoptysis. Seeds are eaten by poorer classes as a substitute of rice or cereal. Young and tender shoots are eaten like *Asparagus* sps. Tender leaves with black pepper and common salt checks diarrhoea in cattle. Leaves and twigs are consumed by elephants. Bamboo is used for house and hut construction.

(Watt, 1889–1893)

Leaves are fed to horses as a remedy for cough and cold. Leaves are cherished by elephants and cattle. Bamboo is used for floating heavy timber and for structural purposes. Split culms are used for basket work etc. Bamboo makes a impenetrable hedge.

(Anonymous, 1948–1976)

Ambasta (1986) prepared uses as in Anonymous (*loc.cit.*).

Stem is useful as medicine on antifertility. Black soot on culms or stem paste is applied on wounds. Leaves and shoots are used as vegetable curry. Stem is also used for house building and basketery.

(Jain, 1991)

Critical Notes

Use of bamboo strips presently noted to prepare traps of fishes or crabs although new, it is not uncommon to employ them for such purposes.

***Bambusa vulgaris* Schrad.

Family: Graminae, Local Name: Bamboo

Ethnobotanical Notes

Miscellaneous

1. Stems are used for house and hut constructions.
2. Three to four feet of stem piece called 'Lodhana' is hanged round the neck and in between forelegs of a cattle to obstruct him from running away.

Critical Notes

Although the above uses are recorded for the first time, the species itself is a substitute for other bamboo species. Other species are widely used as such.

***Barleria lawii* T.And.

Family: Acanthaceae, Local Name: Ekharu

Ethnobotanical Notes

Medicine

1. Plant is eaten as fodder by goats and cattle.

Critical Notes

No other established report is found.

**Bauhinia racemosa* Lam.

Family: Caesalpiniaceae, Local Name: Afta, Sone

Ethnobotanical Notes

Miscellaneous

1. Leaves are used as fodder for domestic animals.
2. Tribal people use leaves for bidi wrapping.
3. Leaves are given to each other as a symbol of love and affection on the eve festival of 'Dasera' especially by Hindu people.
4. Wood is used for building huts.

Notes from Classics

Decoction of leaves is used to relieve head-ache and malarial fevers. Gum is medicinally used. Leaves are pickled by Burmese. Seeds are eaten by the people. Leaves are fed to buffaloes. Leaves are made into cigarette covers or bidies. Tree is a sacred plant of Hindus, worshiped on 'Dasera' Festival.

(Watt, 1889–1893)

Plant yields tannin. Bark yields fibre. It is cultivated as ornamental plant.

(Anonymous, 1948–1976)

Bark is astringent, used in dysentery. Leaves are given with onions on diarrhoea, and as anthelmintic. Decoction of leaves is given in Malaria.

(Ambasta, 1986)

Stem bark is used on dysentery. Stem fibres are used for basketery. Leaves are used for wrapping bidies. Fruits are used as vegetable.

(Jain, 1991)

Critical Notes

Employment of wood for hut construction is unreported yet.

**Bauhinia tomentosa* Linn.

Family: Caesalpiniaceae, Local Name: Bhutcha mul

Ethnobotanical Notes

Medicine

1. Stem pieces kept in 'Taite' and tied around the neck of a person as preventive measure from devil's power.
2. Stem bark pieces are heated and salt is applied on it. These pieces of bark are kept on stomach of a woman for post-pregnancy stomach pains.

Notes from Classics

Decoction of root-bark is useful in inflammation of liver. Dried leaves and young flowers are administered in dysentery. Fruit is diuretic. Seed paste with vinegar is efficacious to wounds inflicted by poisonous animals.

(Watt, 1889–1893)

Decoction of root-bark is used for inflammation of liver and vermifuge. Leaves are used as yellow dye. Dried leaves, buds and flowers are used on dysentery. Fruits are diuretic. Seeds are used as tonic.

(Ambasta, 1986)

Critical Notes

Use of the stem to avoid devils power as well as utility of stem bark to treat stomach pains specially after pregnancy are hitherto unreported.

**Bauhinia vahlii* W&A.

Family: Caesalpiniaceae, Local Name: Chamel

Ethnobotanical Notes

Medicine

1. Bark of this plant and bark of *Mangifera indica* L., *Tectona grandis* L. are crushed and juice is used as for hand-wash five days as a treatment of jaundice.

Veterinary Medicine

1. Bark ash is applied on body of animals to prevent from ticks.

Notes from Classics

Leaves are regarded as demulcent and mucilaginous remedies. Seeds possess tonic and aphrodisiac properties. Seeds are eaten raw.

(Watt, 1889–1893)

Bark yield fibre used for rope making. Leaves are used for making umbrellas. Seeds are eaten raw or fried.

(Anonymous, 1948–1976)

Bark yields cordage fibre. Leaves are used for thatching umbrellas. Seeds are eaten raw or fried like cashew nuts.

(Ambasta, 1986)

Fruits are used in antifertility. Seeds are aphrodisiac, tonic and vermifuge. Plant is useful in dysentery and stomach-ache. Leaves are used for cigarettes and bidis. Fibres are locally used. Fruits are cooked as vegetable.

(Jain, 1991)

Critical Notes

Bark is useful to treat jaundice. It is also employed to prevent ticks. Both applications are not reported in past literature.

**Bauhinia variegata* Linn.

Family: Caesalpiniaceae, Local Name: Shid

Ethnobotanical Notes

Medicine

1. Root paste is applied on dislocations of human body.

Miscellaneous

1. Leaves are used for wrapping bidies.

Notes from Classics

Decoction of root is given in dyspepsia and flatulence. Bark is a alterative tonic and astringent, used on skin diseases, ulcers and to remove intestinal worms. It also prevents decomposition of blood and humours, so used in leprosy and scrofula. Bark, flowers or roots are triturated in rice water as a cataplasm to promote suppuration. Dried buds are used on diarrhoea, dysentery, piles and worms. Buds are eaten as vegetable. Bark is used for dyeing and tanning. Wood is used for agricultural implements, often seen on Buddhist sculptures.

(Watt, 1889–1893)

Bark is described as alterative tonic, useful in scrofula, skin diseases and ulcers. Leaves and buds are edible. Plant is used as goat fodder and also cherished by cattle. Bark is useful for dyeing. Wood is used for making agricultural implements.

(Anonymous, 1948–1976)

Roots are carminative, and their decoction is used in obesity. Stem bark is used as tonic, used against scrofula and cuteneous trouble. Dried flower buds are used in diarrhoea, dysentery and piles. Flowers are laxative, used as pot herb and also pickled. Leaves are used in bidi manufacture. Wood is used for agricultural implements.

(Ambasta, 1986)

Stem bark is used against diarrhoea, dysentery, obesity, leprosy, malaria, bleeding piles, scrofula, skin diseases, ulcers, syphilis and tumours. Leaves and flowers are used to cure worms. Tree is symbolically sacred to Buddha.

(Jain, 1991)

Critical Notes

A new application of roots for dislocations of human body parts is being reported herewith.

**Blumea lacera* (Burm.*f*.) DC.

Family: Compositae, Local Name: Borandu

Ethnobotanical Notes

Medicine

1. Few drops of fresh leaf juice are dropped on wound as disinfectant.
2. About two spoons of decoction of leaves is taken orally against stomach-ache till cure.

Notes from Classics

Fresh roots are chewed as a remedy on mouth dryness. Leaves are astringent and useful for eye wash. Leaf juice is anthelmintic, used for thread-worms and said to be invaluable remedy for *Tinea tarsi*. Plant is regarded as deobstruent and stimulating. It s given in cholera.

(Watt, 1889–1893)

Plant is used as a febrifuge, deobstruent and stimulant.

(Anonymous, 1948–1976)

Plant is used as a febrifuge, diuretic and antiscorbutic. Leaf juice is used as an anthelmintic, particularly against thread worms.

(Ambasta, 1986)

Leaf paste is applied on burns, cuts, wounds, piles, toe injury and in urine complaints. Whole plant is useful for bronchitis and fever. Plants are also helpful for fish killing.

(Jain, 1991)

Critical Notes

The use of fresh leaf juice as disinfectant in case of wounds and decoction of leaves to treat stomach-ache have not been noted in classical literature.

**Blumea malcolmii* Hook. *f.*

Family: Compositae, Local Name: Tantani

Ethnobotanical Notes

Medicine

1. Few drops of leaf juice are dropped in eye at morning against conjunctivitis till cure.

Notes from Classics

Plants yield essential oil.

(Anonymous, 1948–1976)

Critical Notes

Use of leaf juice against conjunctivitis is new to science of ethnobotany.

**Bombyx micaranthes* (L.*f.*) Riedle

Family: Malvaceae, Local Name: Madkey

Ethnobotanical Notes

Medicine

1. Fresh seeds are consumed by native children.

Notes from Classics

Green capsules are edible.

(Ambasta, 1986)

Critical Notes

Consumption particularly of seeds is a new report.

**Borassus flabelliformis* Linn.

Family: Palmae (Araceae), Local Name: Tad

Ethnobotanical Notes

Medicine

1. Roots of *Solanum Virginianum L*. and fruits of this plant are crushed together. Extract is given twice a day for nine days

to increase sex in males and females and as a remedy in impotency.

2. Fruits are cut and applied on ring-worms.

Miscellaneous

1. Sap obtained from plant is taken as drink.
2. Trunk is used as pillars for huts.
3. Leaves are used for thatching roofs.
4. Zip for a Pippet called 'Pipani', and 'Dadari' is made from leaf of Tad.
5. Leaves are used to wrap a musical instrument called 'Pawari'.

Notes from Classics

Root is regarded as cooling and restrorative. Juice expressed from roots and leaf-stalk is used in cases of gastric catarrah and to check hiccup. The fresh juice is diuretic and used in gonorrhoea. Extract of green leaves is used in syphilis. Fresh leaf juice is given as antiplegmatic stimulant in stages of intermittent and remittent fevers. Fresh drawn toddy with rice flour in gently heated and fermentation takes place. This is spread on cloth and applied on part affected to gangarenous ulcerations, carbuncle and indolent ulces. Pulp of ripe fruits is applied externally on skin diseases. Vinegar, toddy and spirituous liquor are made from tree. Slightly fermented juice is used in diabetes. Ash of spadix is given internally in bilious affections. Fresh juice called 'ras' is consumed before sunrise or rapidly ferments and turns milky. Leaves and stalks yield fibres useful for rope making and paper industry. Leaves are used for thatching huts. Leaves are made into 'Pankhas' (Fans), baskets etc.

(Watt, 1889–1893)

Fresh roots are edible. Nira, a refreshing sweet drink is obtained from the plant and is also credited for medicinal virtues. It is used as a stimulant and antiplegmatic and also useful in inflammatory affections and dropsy. Tender shoots are edible. Shoots are ground into flour. Palymra sap is a source of toddy, palymra gur is prepared from sap. Leaf stalks yield fibres, used for brushes and brooms.

(Anonymous, 1948–1976)

Young fresh roots are edible. Tender shoots, fruit pulp, seeds and germinating cotyledons are edible. Leaves were used for writing purposes in ancient times. Leaves are used for making fans, thatching mats, hats, bags, buckets, and also used as green manure and for primitive flutes. Sap is a source of jaggery. Leaf stalks yield fibres used for brushes and brooms. Wood is used for construction, as rafters, pillars, posts and crude bridges. Longitudinally split skin pieces are used as pipes after removal of soft portions. Stem fibres without spinning or twisting are plaited into fish traps.

(Ambasta, 1986)

Leaf petioles are used on hysteria and epilepsy. Wood pulp is considered as abortifacient. Plant is useful on burns, convulsions, ear-ache, head-ache, heat stroke, nausea, menorrhoea, scabies, syphilis, sores and ulcers, also used in gonorrhoea. Inflorescence is used on rheumatism. Leaves are used for making miscellaneous articles, and thatching. Leaves are used as fibre for cordage. Whole plant is carved symbolically on God Vishnu's temple.

(Jain, 1991)

Critical Notes

Medicinal utility of roots to treat impotency is additional and unreported hitherto. Other miscellaneous uses are although new are according to the need and availability of material locally *e.g.*, zip for a trumphet or wrapping 'Pawari'.

**Boswellia serrata* Roxb. ex Colebr.

Family: Burseraceae, Local Name: Salai

Ethnobotanical Notes

Miscellaneous

1. Gum is edible, adulterated with *Acacia* Gum.
2. Wood is used as fuel.

Notes from Classics

Oil distillation from trunk is useful in gonorrhoea, placenta prema, dysmenarrhoea, sore nipple, amenorrhoea and ring worms. Gum as useful to treat rheumatism, nervous diseases, scrofulous affections, urinary disorders, skin diseases, and is generally combined with aromatics. It is regarded as diaphoretic and astringent, used in preparation of ointment for sores, syphilitic diseases, ulcers, diseased bones, buboes etc. Gum mixed with *Acacia* gum is used for foul breath. Flowers and seeds are eaten by Bhils. Timber is recommended for tea boxes. Wood is used as fuel for making charcoal.

(Watt, 1889–1893)

Boswellia oil is used in preparation of varnishes and paints. Timber is used for furniture, co-operage, packing cases, match splints and ammunition boxes. Wood is also used in preparation of mechanical pulp for news print, and plywood etc. Wood is used as good fuel.

(Anonymous, 1948–1976)

Stem bark is used against diarrhoea and also for skin diseases. Gum is considered to be expectorant, diuretic, and stomachic, also used in dysentery, pulmonary affections and cutaneous troubles. Gum and resins are also used in terpentine and varnishes, paints and printing inks, in manufacture of nylon and plastic. Flowers are edible. Wood is used for paper making, plywood manufacture, veneers, furniture, packing cases, toys, ammunition boxes and for carvings.

(Ambasta, 1986)

Stem bark is antiseptic, used on cold, cough and tooth-ache. Leaves are used in the treatment of eye diseases. Resins are used to control diarrhoea, dysentery, pulmonary diseases, skin eruptions, sores and stomach complaints. Plant is useful in cases of epilepsy. Seeds are used for better hair growth. Resins are used as veterinary medicine. Seeds are used as antidote on snake-bite. Wood is used as fuel for smelting iron. Poles of wood are auspicious for wedding place. Ample of flowers indicate good crop of cereals.

(Jain, 1991)

Critical Notes

Gum is reported medicinally useful, however, its edibility is not reported.

**Brachiara reptans* (Linn.) Gard. & C.E.Hubbard

Family: Graminae, Local Name: Chopadi

Ethnobotanical Notes

Medicine

1. Leaf extract, about half cup daily for a fortnight, is given for anaemic person.

Notes from Classics

Grains are eaten at times of scarcity. Plant is used as cattle fodder.

(Ambasta, 1986)

Critical Notes

Usefulness of leaf extract to treat anaemic condition has remained hitherto unreported.

**Bridellia airy-shawii* P.T. Li

Family: Euphorbiaceae, Local Name: Aasand

Ethnobotanical Notes

Medicine

1. Fresh stem bark is heated and juice tapped from it is poured in ears to treat ear-ache.

Miscellaneous

1. Food: Ripe fruits are eaten raw.

Notes from Classics

Bark oil is used as a liniment with gingelly oil in rheumatism. Bark is useful for tanning. Wood is used for making yoke and agricultural implements, carts etc. Leaves are fed to cattle to prevent from worms.

(Watt, 1889–1893)

Leaves are served to cattle as fodder. Bark is used for tanning purposes. Timber is used as second class wood for constructional work, making carts, and agricultural implements. Fruits are edible.

(Anonymous, 1948–1976)

Ambasta (1986) reported uses similar to those in Anonymous (*loc.cit.*).

Bark is used as veterinary medicine. Leaves are used as cattle fodder. Bark is also used as dye for coarse wood.

(Jain, 1991)

Critical Notes

Fruits are already reported for medicinal utility, however, their consumption as food is new. Stem bark is being reported useful to treat earache. It is new application.

**Butea monosperma* (Lam.) Kuntze

Family: Papilionaceae, Local Name: Palas

Ethnobotanical Notes

Medicine

1. Extract of heated roots, about two-three spoons, is given at night as a remedy for impotency. It is advised for a month.
2. Stem bark is used to apply on injury caused due to axe.
3. Stem juice is applied on goitre of humans being.
4. Flowers are useful in spermatorrhoea, and to prevent pus from urinogenital tract of males.
5. Seeds are consumed by children as a remedy for intestinal and stomach worms.

Miscellaneous

1. Leaves are used for making 'Ghongada' to protect from rains in rainy days.
2. Leaves are weaved and used a dinning plates and bowels.
3. Twigs are tied on horns of bullocks on occasion of 'Pola' festival.

Notes from Classics

Leaves are astringent, tonic, aphrodisiac and are used to disperse boils, and pimples, given internally in flatulent colic, worms, and piles. Leaves are used as fodder for buffaloes, and elephants, also regarded as a valuable manure. Leaves are used as platters. Dried twigs are used for feeding of 'hom' or sacred fire in the name of 'nava grahas' on the occasion of 'vastu-shanti'. Flowers are astringent, depurative, diuretic and aphrodisiac. Poultice of flowers is used to disperse swellings and promotes diuresis and an menstrual flow. They are given to pregnant women in cases of diarrhoea, and externally in orchits. Seeds are purgative, largely used in the treatment of round-worms. Seeds pounded with lemon-juice and applied to the skin, act as rubefacient. Fresh juice is used in pthisis and haemorrhagic affections. It is employed in ulcers and relaxed sore throat. It is given in diarrhoea and dysentery. Gum is astringent. Tree is sacred to 'Soma' (Moon). Wood is sacrificial and mentioned in Vedas.

(Watt, 1889–1893)

Roots yield fibre and used for rope making. Leaves are much used for platters, cups etc. Seed powder is a good remedy for ascaris, round-worms, and tape-worms. Wood is used to prepare well-crubs and scoops etc. It is also used as fuel. Dried twigs are used to fed the sacred fire.

(Anonymous, 1948–1976)

Root bark fibres are used for cordage and caulking boats. Stem bark is astringent and used for piles, tumours and menstrual disorders. Seeds are pounded with lemon-juice which acts as a powerful rubefacient and are used to cure for a form of herpes called 'Dhobies' itch. Gum is having astringent properties, used in diarrhoea. Wood is useful for making well-curbs, and water scoops, structural work and also useful for news print manufacture.

(Ambasta, 1986)

Roots are useful in snake-bite and tuberculosis. Stem bark, flowers and gum are used medicinally on diarrhoea, dysentery, haemost and applied on wounds. Stem bark and flowers are used as fish-poison. Bark yields dye and fibre. Leaf paste is applied on bone fracture, boils, delirium, inflammation, skin diseases and vermifuge. Flowers are useful in antifertility, diabetic, eye complaints, ascaris, piles, pimples, stomach-ache, sunstroke, and cures urine complaints, also used against dog-bite. Seeds cure abortificient, antihelmintic, contraceptive and said to be cooling. Gum is useful as tonic. Plant is a good remedy for menorrhoea. Flowers are used by natives as vegetable.

(Jain, 1991)

Critical Notes

All uses related to roots, stem, leaves and flowers are new applications of this species.

**Butea superba* Roxb.

Family: Papilionaceae, Local Name: Palasvel, Vadha

Ethnobotanical Notes

Veterinary Medicine

1. Tuberous roots are heated and applied on inflammation and swellings in case of cows or buffaloes.

Miscellaneous

1. Fibres are pulled out from stem to tie fodder and fire wood bundles.
2. Entire lianas are collected without cutting them into pieces. They are used for huts to fasten various things employed for its construction.

Notes from Classics

Stem yields fibres used for tying grass or firewood bundles. Leaves are used as fodder for cattle.

(Watt, 1889–1893)

Roots and young branches yield fibres, used for rope making. Leaves are eaten by cattle.

(Anonymous, 1948–1976)

Decoction of shoot is considered as emollient and sedative, used on piles. Seeds are anthelmintic and yield fatty oil. Leaves are used in tonic and poultice.

(Ambasta, 1986)

Stem bark is applied on cuts. Plant is useful in dysuria and night fever. Flowers are useful for antifertility. Roots are applied on neck ulcers of cattle.

(Jain, 1991)

Critical Notes

Use of tuberous roots for veterinary medicine is not yet noted.

Caesalpinia decapetala (Roth) Alst.

Family: Caesalpiniaceae, Local Name: Chilar

Ethnobotanical Notes

Miscellaneous

1. It is planted around cultivated fields as live hedge.
2. Leaves are cherished by goats.

Notes from Classics

Bark is used for tanning. Leaves are bruished and applied on burns. Plant forms impenetrable fence.

(Watt, 1889–1893)

Roots are used on rheumatism. Tender shoots are bruished and used in opthalmia. Leaves are used as fodder. Wood is used for cog-wheels, walking sticks and tent-pegs. Bark yield fibre.

(Anonymous, 1948–1976)

Roots are purgative. Bark is used for tannin.

(Ambasta, 1986)

Entire plant is used as a remedy for jaundice. Leaves are given to cattle for sore mouth.

(Jain, 1991)

**Cajanus cajan* (Linn.) Millsp.

Family: Papilionaceae, Local Name: Tur, Tor

Ethnobotanical Notes

Medicine

1. Leaves of this plant and stem bark of Umber (*Ficus racemosa*

Linn.) are crushed. This extract is given to a wine drinker to reduce effect of alcohol.

Miscellaneous

1. Stalks are tied in bundles and are used for cleaning food-grains on farmyards.
2. They are used to clean fore ground.
3. Stalks are also employed for making partition walls of huts.
4. Dried stalks are used for fuel.

Notes from Classics

Leaves are used on mouth diseases. Paste of leaves and the pulse is warmed and applied over breast to check secretion of milk. Tender leaves are chewed in cases of apthae and spongy gums. Poultice made with seeds will check swellings. This is third rank leguminous seed most commonly used in vegetable curry. Sometimes cooked green. Sweet cakes are made from seed flour. Leaves are good fodder for cattle. Wood is used as fuel. Stalks are used for making baskets. Stalks are also employed in gunpowder works for charcoal.

(Watt, 1889–1893)

Dried stems are used for thatching huts and also used as fuel. The husk of pods and seeds is used as cattle feed mixed with broken dal called 'chunni', used for milking animals. Seeds are extensively proteinous, diet for mankind. Leaves are used as fodder and green manure.

(Anonymous, 1948–1976)

Pods are cooked and eaten as vegetable. Seeds are extensively eaten in regular diet. Leaves are fed to silkworms and also used as fodder. Leaves are considered to be a good manure.

(Ambasta, 1986)

Leaves are used for gargle, on skin burns, sores on mouth and tongue. Seeds are eaten as a pulse or vegetable.

(Jain, 1991)

Critical Notes

To bar the effect of alcohol, leaf extract is employed. This use did not find place in the classical literature. The species is cultivated for pulse and has also wide miscellaneous uses. It is exotic, native of Africa.

**Calotropis gigantea* (Linn.) R.Br.ex Ait.

Family: Asclepiadaceae, Local Name: Ruchkin, Mhatari

Ethnobotanical Notes

Veterinary Medicine

1. Latex is dropped and pasted on shoulder of a cattle to prevent from 'Khargoti', a nervine disease which hinders walking of an animal.
2. To cure swelling on shoulder of a bullock, latex is applied below tail of a bullock. A warm knife is kept on it to cure.

Miscellaneous

1. A garland of leaves and flowers is placed around the neck of an idol of God Shani and Maruti by Hindus.

Notes from Classics

Root bark is useful in skin diseases enlargement of abdominal viscera, intestinal worms, cough, ascites and anasarca. Paste of root bark with sour congi, is applied on elephantiasis of the leg and scrotum. Leaves are mixed with rock salt, are roasted in closed vessel so that fumes may not escape. The ashes thus produced are given with whey in ascites and enlargement of the abdominal viscera. Fresh leaves are roasted and used for joint pains and swellings. Leaves are boiled in coconut oil and this oil is applied on paralysis. Powder of dried leaves is dusted on wounds to destroy excessive granulation and promote healthy action. Fresh flowers in equal proportion of black pepper is useful and cheap remedy for asthma, hysteria and epilepsy. Flowers are digestive, stomachic and tonic, useful on cough, asthma, catarrah and loss of appetite. Drug is prescribed in leprosy. Bark is described as alterative tonic, diaphoretic and large doses are emetic. Drug is also useful for syphilis, diarrhoea, dysentery and chronic rheumatism. Milky juice is regarded as drastic purgative and caustic, generally used in combination with latex of *Euphorbia nerilifolia*. Drug is irritant poison, large doses produces vomiting. Sap mixed with flour of small millet is used previously colouring the skin with lac dye.

(Watt, 1889–1893)

Root bark powder is useful on dysentery and elephantiasis. A tincture of leaves is used on the treatment of intermittent fever. Flower powder in small doses is used in the treatment of cold, cough, asthma and indigestion. Latex is strong irritant to skin and mucous membrane. Latex is used in tanning industry. It is adulterated with persian opium. Stem bark yields fibre, used for making fishing nets, lines, strings and twine. Soft and glossy floss of seeds is used for stuffing mattresses. Wood yields light charcoal used in gunpowder and fire works.

(Anonymous, 1948–1976)

Leaves are rubbed on skin of elephants as a treatment of 'kesaryer' disease. Stem bark yields fibre, used for fishing nets, bow strings and twine. Seeds yield fixed oil. Floss from seeds is used for stuffing pillows and mattresses.

(Ambasta, 1986)

Root bark is useful to control dysentery, post-natal and guinea worms. Bark is used on child-birth and epilepsy. Leaves are considered to be useful on body-ache, injuries ring worms and skin diseases. Leaf juice is dropped in ears for ear complaints and eye complaints. Latex is used on hydrophobia, leprosy, stomach-ache, tooth-ache, worms in gums and wounds. Plant is abortificient, used on anasarca, boils, carbuncle, burns, paralysis of tongue, scabies, small pox, syphilis, and as tetanus. Plant is symbolically worshipped. The knotted roots are regarded abode of Lord Shiva and Ganesha. Sulptures and temples show that 'Parvati' hides in its flowers to save herself from defy demons. Bachelor should marry the plant before marrying widow.

(Jain, 1991)

Critical Notes

Application of latex for a nervine 'Khargoti' disease, also for swelling on shoulder are not documented yet. Garlanding by the wreaths of leaves of this species is common method of worshipping God Maruti in this region.

**Calotropis procera* (Ait.) R.Br.

Family: Asclepiadaceae, Local Name: Rui

Ethnobotanical Notes

Medicine

1. Roots are loosely fastened around stomach of a pregnant woman for easy delivery.
2. Paste of root bark alongwith lime and catechu is chewed and swallowed to cure stomach-ache.
3. Tender leaves are chewed in cold.
4. Latex is applied on forehead for seven days as a remedy against catarrah.

Veterinary Medicine

1. Twenty one terminal leaf buds of this plant and *Clematis triloba* are crushed with water and slurry is applied on infected part of a cattle. First day of the treatment should be Saturday and the animal is not allowed for any agricultural work for about 7–8 days.
2. Latex is applied on dorsal basal part of tail of pet animals to cure shoulder injury.
3. Latex is applied on skin infection called 'Khargoti' (swelling) on shoulder of cattle.

Notes from Classics

Roots are used as toothbrush as a remedy for tooth-ache and for guinea worms. Mouth of uterus is brushed with fresh twigs in some parts of India. Flowers are used on cholera. Fresh latex is used as infanticide in Punjab. Plant is useful as camel fodder.

(Watt, 1889–1893)

Stem yields fibre similar to that of *Calotropis giagntea*.

(Anonymous, 1948–1976)

Root bark is used in leprosy. Stem yields fibre. Floss is used for stuffing mattress and pillows. Wood yields gunpowder and charcoal.

(Ambasta, 1986)

Roots are useful on asthma, cough, fever and swelling. Leaves are used against boils, cold, eczema, skin diseases, stomach complaints and swellings of mouth. Flowers are useful in cholera. Plant is useful on epilepsy, small-pox, sores and spleen complaints. Roots are used as veterinary medicine. Latex is abortifcient, used as antiseptic, also used on dropsy, rheumatism, ringworms, tooth-ache, taenia, wounds and masticatory. Latex is used as antidote on snake bite and scorpion sting. Floss is used as fibres.

(Jain, 1991)

Critical Notes

Utilities of roots for easy delivery, its bark to treat stomach-ache and latex against cattarrah are new to the science of ethnobotany. All the three veterinary uses noted by authors are likewise new to the science.

**Canavalia virosa* (Roxb.) Wight & Arn.

Family: Papilionaceae, Local Name: Ran-abhay

Ethnobotanical Notes

Medicine

1. Root paste is mixed with latex of *Ficus arnottiana* (Miq.) Miq. The paste is applied on swelling of thigh, called "Mandi Pata".

Notes from Classics

Green pods and beans are cooked and eaten as vegetable.

(Ambasta, 1986)

Critical Notes

Curing swelling of thigh by the root paste is a new medicinal use. The plant is not reported earlier as useful for medicinal purposes.

**Canscora diffusa* (Vahl) R.Br.

Family: Gentianaceae, Local Name: Mayaco, Zinku, Bhuinarali

Ethnobotanical Notes

Medicine

1. Leaves are burnt and 'Misri' (ash) is mixed with coconut oil and applied on skin to prevent from hair fall because of skin infection.
2. Flowers of this plant and roots of *Brassica oleracea* L. are crushed in milk. This mixture about one spoonful is given twice a day for three days to reduce fever of children.
3. Entire plant is burnt and the smoke is passed over body as a remedy on swellings.

Veterinary Medicine

1. Plant is burnt and 'Misri' (ash) of it is applied on swellings, inflammations and bruises of cattle.

Notes from Classics

Plant is regarded as laxative, alternative and tonic, much praised as nervine tonic, used in epilepsy and nervous debility. Fresh juice is given in all cases of insanity. Paste of entire plant taken with milk is considered as a nervine and alterative tonic.

(Watt, 1889–1893)

Plant is pungent and oleaginous and is regarded laxative alterative, and tonic. Fresh juice of plant is prescribed in insanity, epilepsy and nervous elebility.

(Anonymous, 1948–1976)

Leaves are laxative. Whole plant is used on head-ache, malaria and fever. It is symbolically used on religious ceremony.

(Jain, 1991)

Critical Notes

Use of leaves for falling of hair, of flowers to reduce fever in children and entire plants on swellings have not appeared in classical literature. Similarly, application of plant ash to treat swelling, inflammation and bruises in case of cattle are newly reported.

**Capparis sepiaria* L.

Family: Capparidaceae, Local Name: Pilukatar

Ethnobotanical Notes

Medicine

1. Roots are crushed in water. The extract is applied on swellings.
2. Fruits are fried and the epicarp is consumed for stomach-ache.

Miscellaneous

1. Food: Ripe fruits are eaten. Unripe fruits are cooked as vegetable.

Notes from Classics

Plants possess febrifuge properties, hence recommended in fevers. It is also considered as alterative and tonic, and useful in skin diseases. Branches are used for hedges.

(Watt, 1889–1893)

Plant is considered as febrifuge, alterative and tonic, and also useful in skin diseases. Plant is also useful as hedge plant.

(Anonymous, 1948–1976)

Plant is having febrifugal and tonic properties. Plant is useful for skin complaints.

(Ambasta, 1986)

Roots and bark are used on dropsy and gout. Roots are used in ear-ache and mumps. Leaves are used on skin diseases. Whole plant is used on swelling.

(Jain, 1991)

Critical Notes

Medicinal utility of fruits against stomach-ache is unnoticed. Edibility of fruits is also not on past records.

**Capparis spinosa* L.

Family: Capparidaceae, Local Name: Waghati

Ethnobotanical Notes

Medicine

1. Decoction of bark two spoon per day is used on rheumatism till cure.
2. Fodder: Plant is used as goat fodder.

Notes from Classics

Root bark is hot and dry, acts as detergent and astringent, expels cold humours and therefore recommended in plasy, dropsy, gouty and rheumatic actions. Dried stem bark is diuretic and employed on obstructions of liver and spleen, amenorrhoea and rheumatism. Plant juice is dropped in ears to kill worms. Flower buds or flowers treated with salt and vinegar. They are pickled which has burning taste. They are considered to be useful in scurvy. Ripe fruits are eaten or made into pickles. Leaves are eaten as spinach. Leaves and ripe fruits are cherished by sheep and goats.

(Watt, 1889–1893)

Stem bark is diuretic, expectorant and tonic, used in affections of liver, rheumatism, tubercular glands and paralysis.

(Ambasta, 1986)

Bark is used as a remedy on paralysis and tooth-ache. Bark and roots in combination are used on rheumatism. Fruits are edible. Flower buds and leaves are cooked as vegetable.

(Jain, 1991)

Critical Notes

Administration of bar decoction is reported useful for the first time on rheumatism.

Capparis zeylanica L.

Family: Capparidaceae, Local Name: Waghati, Govindi

Ethnobotanical Notes

Miscellaneous

1. Fruits are cooked and eaten as vegetable.

Notes from Classics

Green fruits are pickled and eaten.

(Watt, 1889–1893)

Root bark is used in cholera. Leaves are used as a counter-irritant and as a cataplasm in boils, swellings and piles. Green fruits are used on piles.

(Anonymous, 1948–1976)

Root bark is sedatine and stomachic, and also used in cholera. Fruits are pickled and eaten.

(Ambasta, 1986)

Root bark is anthelmintic used for boils and cholera. Root bark is used for swellings of testicles. Leaves are used against piles, and swellings. Fruits and roots are used as antidote on snake-bite. Fruits are pickled and also used as vegetable.

(Jain, 1991)

**Capsicum annuum* L.

Family: Solanaceae, Local Name: Lawangi mirch, Mirchi

Ethnobotanical Notes

Medicine

1. Five stem pieces are fastened on ear opening to get relief from migraine. This is followed for a week.

Miscellaneous

1. Fruits are used as spice in curries with vegetables.

Notes from Classics

Drug is considered to be stomachic and stimulant, hence externally used as rubefacient. Fruit drug with sugar and tragacanth are remedy for hoarseness with professional singers and public speakers. Fruits are valuable in cold, for leucophlegmatic habits, dyspepsia and flatulence. Fruits are used as stimulant on snake-bite. Fruits when green are pickled and also eaten raw or with sauces, curries, chutneys etc.

(Watt, 1889–1893)

Capsicum preparations are used as counter irritants in lumbago, neuralgia, and rheumatic disorders. It is internally taken as carminative action, useful in dyspepsia. Large quantity of dry and green chillies are consumed as a spice.

(Anonymous, 1948–1976)

Ambasta (1986) recorded uses similar to those in Anonymous (*loc.cit.*).

Fruits are eaten as condiment, eaten with vegetable or in curries.

(Jain, 1991)

Critical Notes

Stem is thought useful in case of migraine. This use is not mentioned in the classical literature. It is cultivated as a crop for its pungent fruits. It is exotic, native of Tropical America.

Caralluma adscendens (Roxb.) R.Br.

Family: Asclepiadaceae, Local Name: Vanjari Bhaji

Ethnobotanical Notes

Miscellaneous

1. Plant is cooked and eaten as vegetable.

Notes from Classics

Entire fleshy plant is eaten by natives in the form of prickles or made into chutney.

(Watt, 1889–1893)

Plant is used as vegetable.

(Anonymous, 1948–1976)

Plant is used as vegetable.

(Ambasta, 1986)

Plant is used as vegetable.

(Jain, 1991)

Critical Notes

The classical literature and the present use as vegetable amply indicate better food value of this species.

Cardiospermum halicacabum L.

Family: Sapindaceae, Local Name: Kapal-phodi, Bhaura

Ethnobotanical Notes

Medicine

1. Few drops of leaf juice are poured in ears to treat ear-ache.
2. Entire plant is used as cattle fodder.

Notes from Classics

Roots are used as an emetic, rubefacient, laxative and stomachic. It also possess diaphoeretic, diuretic and tonic properties. It is prescribed on rheumatism, nervous complaints and piles. Fried leaves bring out secretion of menses. Leaves mixed with castor oil is externally employed in rheumatism and lumbago. Leaves mixed with jaggery and boiled in oil are good in sore eyes. Entire plant is boiled in oil and employed to anoint the body in bilious affection. Juice of plant, one spoon daily, promotes catamenial flow during menstrual period and gonorrhoea.

(Watt, 1889–1893)

Roots are diuretic, diaphoretic, emetic, laxatives, rubefacient and emmenogogue. They are occasionally used for rheumatism, lumbago and nervous diseases. Leaf juice is useful to cure ear-ache.

(Anonymous, 1948–1976)

Ambasta (1986) mentioned uses similar to those in Anonymous (*loc.cit.*).

Roots are useful of dizziness. Stem is used on fever. Leaves are used on wounds to reduce bulkiness, on ear-ache and anasarca. Leaves roots and seeds together are used on eye sores, ear-ache, lung diseases, rheumatism and stiffiness of limbs. Plant is used on nervous disorders.

(Jain, 1991)

**Careya arborea* Roxb.

Family: Lecythidaceae, Local Name: Kumbhai

Ethnobotanical Notes

Medicine

1. Stem bark extract, about 20–25 ml, is administered orally to get relief from piles. It is advised till cure.
2. Bark pieces are heated and kept on annul opening to treat piles.

Miscellaneous

1. Fruits are eaten by livestock.

Notes from Classics

Stem bark is used on wounds. Flowers and bark juice mixed with honey is used as demulcent, in cough and cold. Infusion of flowers is given by native-midwives to heal ruptures caused by child-birth. 'Sherbet' prepared from flowers is tonic after child-birth. Fruit decoction is given to promote digestion. Infusion of stem bark is given as an antidote on snake bite. Bark is used for tanning. Fibrous bark is used as slow match to ignite gun powder in Mysore. Fibre is good source of cordage. Wood is used for preparing agricultural implements, cabinets etc. Seeds are said to be poisonous.

(Watt, 1889–1893)

Bark is said to be demulcent and so recommended on cough and cold. The inner bark rubbed on shoes to keep of leaches. Leaves are used for ulcers. Fruits are aromatic and edible. Decoction of fruit is given to promote digestion. Bark is used for making 'Brown' paper and coarse cordage. Timber is useful for making agricultural implements cabinets, gun shocks, house-posts, planks and railway sleepers.

(Anonymous, 1948–1976)

Ambasta (1986) described uses similar to those in Anonymous (*loc.cit.*).

Stem bark is used against blood dysentery, cuts, wounds, eye diseases and skin diseases. Leaves and bark are used on cold and cough, also used on swelling faces. Fibre is used for making cordage.

(Jain, 1991)

Critical Notes

Utility of bark against piles is not reported earlier. Consumption of fruits by livestock is also an additional use of this taxon.

**Carissa congesta* Wight.

Family: Apocynaceae, Local Name: Karwand

Ethnobotanical Notes

Medicine

1. Root extract is administered for 1–2 days to control vomiting.
2. Roots of this plant and inflorescence of *Zea mays* L. are boiled together in water and cooled. A cup of this decoction for 15 days is taken orally in the treatment of kidney stone.
3. Latex is dropped on urethral opening against burning sensation during urination.

Veterinary Medicine

1. Red lead is applied on roots, then are tied around the neck to cure wounds and septics.

Miscellaneous

1. Ripe fruits are edible.
2. Ripe fruits are said to be useful for preparing alcoholic beverages.
3. Branches are used as hedge for fields and farm-yards.
4. Branches are used for 'Rabani'. In which they are burnt in field before sowing Nagali seeds (*Eluesine coracana* Gaertn.).

Notes from Classics

Roots are stomachic, used as plaster to keep of flies. Roots are powdered with camphor and lime juice. It is applied as a remedy for itching. Decoction of leaves is used in commencement of remittent fever. Unripe fruit is astringent, and ripe ones are cooling and hence useful in bilious complaints. Fruits are considered to be antiscorbutic, and much used in curry, chutney, pickles, puddings etc. Jelly can be prepared from ripe fruits. Plant is useful for fencing.

(Watt, 1889–1893)

Decoction of leaves is given in early stages of remittent fever. Unripe fruit is used for pickles, ripe fruits are edible, suitable for tarts, puddings and jellies. Wood is used for making combs and spoons. Plant is suitable for fencing.

(Anonymous, 1946–1976)

Roots are considered stomachic and anthelmintic. Decoction of leaves is given in remittent fevers. Ripe fruits are edible, and suitable for tarts, puddings, jellies and unriped fruits are pickled. Tree in planted for fencing 'Tasser' silk worms. These are also fed on it.

(Ambasta, 1986)

Leaves are useful on fever. Seeds are used against itching. Plant is used on anasaraca, dropsy, epilepsy, madness, post-natal care, rheumatism and to cure sores. Root bark is used to cure wounds and worms from cattle.

(Jain, 1991)

Critical Notes

Roots to control vomiting and against kidney stones, wounds and septic, and latex for burning sensation during urination are not reported in past. To prepare beverage, ripe fruits are also employed. This is also not recorded earlier.

**Carum carvi* Linn.

Family: Umbelliferae, Local Name: Jire

Ethnobotanical Notes

Veterinary Medicine

1. Fruits of *Carum carvi* Linn. (in larger proportion) *Syzygium aromaticum* (Linn.) Merrill & Perry and *Piper nigrum* Linn. are mixed with bark of *Cinamomum zeylanicum* Bth. and ground together. This powder is homogenized with ghee (of vegetable origin) and one egg of hen. It is then given to animals for tongue infections and sores in mouth. It is also given in the treatment of flatulence.

Notes from Classics

Dried fruits possesses stimulant and carminative properties useful in flatulent colic, atonic dyspepsia, and spasmodic affections of bowels. Fruit oil is used as stimulant. Fruits are considered as aromatic, carminative and astringent. A caraway bath is recommended for painful swelling of womb and poultice is useful for painful and protruding piles. Seeds are employed to flavour curries, cakes and confectionary drinks.

(Watt, 1889–1893)

Roots are eaten as vegetable. Dried fruits are widely used as spice and for flovouring bread, meat, sauces, vegetable, in cheese, brandy and liquour. Fruits are used as mild stomachic, and carminative in flatulent colic, contains and essential oil, used for flavouring and as carminative.

(Ambasta, 1986)

Seeds are carminative, used in cold, cough, dyspepsia, fever, rheumatism and stomach disorders. Plant is useful in gonorrhoea. Entire plant is edible.

(Jain, 1991)

Critical Notes

The plant is being reported useful to treat veterinary diseases for the first time. Although it is a cultivated species, but is not usually cultivated in this district.

Carvia callosa (Nees) Bremek.

Family: Acanthaceae, Local Name: Carvi

Ethnobotanical Notes

Miscellaneous

1. Stems are used for thatching walls as well as roofs.
2. Leaves used to prepare beds for onion storage in sheds.
3. Dried branches are used as hedge for farmyards.

Notes from Classics

Bark is emollient. Flowers are vulnerary. Source of lime pulps is used for making cheap wrapping papers, flowers are also good source of honey. (Ambasta, 1986)

Critical Notes

Miscellaneous use for storage of onions noted by the authors have still remained unnoticed.

Casearia esculenta Roxb.

Family: Flacourtiaceae, Local Name: Kirmira

Ethnobotanical Notes

Miscellaneous

1. Stem pieces are kept in pockets or caps by the tribals. This, in their opinion, helps to keep away the poisonous reptiles during their stay in forest areas.

Notes from Classics

Roots are purgative and used by people of hilly region. Leaves are eaten in stews by natives.

(Watt, 1889–1893)

Root is astringent and purgative. Decoction of roots is used in diabetes and piles. Leaves are used in medicated baths. Leaves are eaten in stews. Fruits are edible.

(Anonymous, 1948–1976)

Uses recorded by (Ambasta, 1986) are similar to those uses of Watt (*loc.cit.*).

Critical Notes

The use noted by present investigator is not noted earlier.

Cassia fistula L.

Family: Caesalpiniaceae, Local Name: Bahava

Ethnobotanical Notes

Medicine

1. Leaf juice is dropped in ears twice a day for three days in ear complaints.
2. Infusion of inner fruit pulp is administered to infants reluctant to breast feeding.
3. Inner fruit pulp is chewed as a remedy on stomach ache.
4. Mature fruits are crushed in water, this paste is applied on body and bathed after one hour as preventive measures from devil's powder.

Veterinary Medicine

1. Five stem pieces of this plant L. and of *Emblica officinalis* Gaertn. are tied together. Red lead is applied alongwith fibre obtained from *Hibiscus canabinus* L. and 'Dhup' is given on burning charcoal adding edible oil on charcoal. Fibres are passed five times over body of a cattle from head to tail and tied around neck of cattle to repel worms from wounds.

Miscellaneous

1. Stem pieces of this plant and *Cassine albens* (Retz.) Kosterm. are kept in crop fields of rice (*Oryza sativa* L.) and Nagali (*Eleusine coracana* Gaerth.) for better yield and believed that it protects these from devil's power.
2. Pulp of pod is liked by the tribal children.
3. Dolls or Toys are carved from wood.
4. It is believed that if a person crosses cover the mature pod dropped on ground, becomes blind.

Notes from Classics

Bark combined with tamarind is externally useful on rheumatism and gout. Bark and leaves are rubbed with oil and applied on pustules, skin diseases and ring-worms. Infusion of leaves is laxative. Poultice made from leaves is said to be beneficial if rubbed on affected part of facial paralysis and rheumatism, internally also given on brain affections. Leaf paste is useful on ringworms. Flower confection made into 'Gulkand' is used as a febrifuge, also useful in cases of diabetes and mellitus. 'Gulkand' is cooling, laxative used in constipation in delicate women, ½ once with warm milk is given at bed time. Ripe pods of same plant and pods of tamarind are made into pulp and taken as a drink at night, acts on bowels mildly. In flatulent and colic, pulp is commonly applied around the naval area to produce motions. Pods are warmed and pulp is extracted, which is rubbed with almond oil. It is used as a safe purgative for children and pregnant women. It is also prescribed as linitive and useful in relieving thorasic obstructions. Tender leaves are eaten with food by natives. Wood is used for making agricultural implements carts and rice pounders.

(Watt, 1889–1893)

Root bark extract is used in the treatment of black-water fever. Wood is used for houses and bridge posts, agricultural implements, ploughs, tools, carts, etc.

(Anonymous, 1948–1976)

Root bark extract is used to treat black-water fever. Stem bark mixed with *Avarum* bark is used for tanning. Dried fruits are purgative and laxative, used for habitual constipation. Wood is used for making agricultural implements, carts and rice pounders.

(Ambasta, 1986)

Roots are used on skin diseases and swelling of throat. Stem bark is useful on blindness, diarrhoea and also useful in the treatment of dog-bite. Leaves are used as antiseptic, prescribed on blood purification, dropsical swelling, rheumatism and ring-worms. Fruits are used against chest infections, constipation, cough, diarrhoea, dysentery, indigestion, liver complaint, pimples, swellings and tooth-ache. Seeds are used as cooling agent, recommended as gargle in jaundice, gastric complaints and as vermicide. Plant is useful on burns, dysuria, epilepsy, syphilis. Wood is used for house construction. Tree is symbolically abode of god. Bark and fruits are used against scorpion sting and as an antidote on snake-bite. Flowers and fruits are eaten as vegetable.

(Jain, 1991)

Critical Notes

Leaf juice is beneficial for ear complaints. Fruits are useful in case a child is unwilling for breast feeding. Likewise, they are useful for stomach-ache. These uses are not recorded yet. There are also some ethnobeliefs described above which are also unpublished.

**Cassia occidentalis* L.

Family: Caesalpiniaceae, Local Name: Ran-tarota

Ethnobotanical Notes

Medicine

1. Roots of this species along with roots of *Caesalpinia sepiaria* Roxb. and *Azadirachta indica* A. Juss. are kept in water. This infusion is given against white discharge in ladies. About 20 ml, is administered daily at night till cure.

2. Leaves are burnt and soot obtained is mixed with coconut oil and applied on eye-lids for cooling sleep.

Notes from Classics

Infusion of roots is considered beneficial in obstruction of stomach and incipient dropsy. Roots are used as diuretic in West Indies. Leaf paste is externally applied on healing wounds, sores, itch and cutaneous diseases. Mature seeds are used on ring worms. Seeds are used as Negro Coffee also used as febrifuge. Roots are used as veterinary medicines for animal diseases. Infusion of roots is considered as antidote in cases of poison.

(Watt, 1889–1893)

Seeds are externally applied on skin diseases. Seeds are used as substitute for coffee. Plant is used as green manure.

(Anonymous, 1948–1976)

Ambasta (1986) mentioned uses similar to those in Anonymous (*loc.cit.*).

Roots are used against gastric complaints, to increase lactation, used on whooping cough in combination with seeds. Leaves are used on bone fracture, fever, ring-worm, skin diseases, throat infection and wounds. Twigs are used as tooth brushes. Plant is used against diarrhoea, dysentery and rheumatism. Seeds are used as stimulant drink. Plant is used as antidote on snake-bite. Leaves are useful on veterinary wounds.

(Jain, 1991)

Critical Notes

Roots against white discharge in ladies and leaves for cool sleep are new applications. It is exotic weed of waste lands, native of South America.

**Cassia tora* Linn.

Family: Caesalpiniaceae, Local Name: Tarota

Ethnobotanical Notes

Medicine

1. Leaves are crushed and paste is rubbed on body for curing fever.

Miscellaneous

1. Seeds are ground and used as a feed for bullocks.

Notes from Classics

Roots rubbed with lime-juice is used to cure ring-worm. Leaves and seeds constitute valuable remedy on skin diseases and itch. Seeds are steeped in juice of *Euphorbia nerilifolia* L. and cow urine, their paste is applied for cheloid tumours. Seeds are substituted to coffee seeds. Tender leaves are boiled and cooked as pot-berb.

(Watt, 1889–1893)

Leaves are useful on skin diseases. Seeds are roasted and used as substitute for coffee. Seeds are laxative. Seeds are fed to cattle.

(Anonymous, 1948–1976)

Leaves are purgative, used on ring-worms and other skin diseases. Seeds are used as a substitute for coffee and as mordant in dyeing.

(Ambasta, 1986)

Roots are used on abnormality of child birth, also help in fever and ring-worms. Leaves are used as antiseptic, on boils, cuts, bone fracture, indigestion, intestinal diseases, jaundice, night blindness, rheumatism, stomach ache, wounds, and also used as vermicide. Seeds are used as galactogogue and tonic to increase fats, cord and eczema. Entire plant is used on epilepsy and scabies. Flowers are eaten as vegetable.

(Jain, 1991)

Critical Notes

Medicinal utility of leaves to cure fever is unrecorded yet. The plant species is exotic, native of America.

**Cassine albens* (Retz.) Koesterm.

Family: Celastraceae, Local Name: Buscut, Bhutya Kalas

Ethnobotanical Notes

Medicine

1. Leaf powder is snuffed, which gives relief from cough and head-ache.

2. Latex or gum is applied on lips to cure cracks.
3. Leaf juice, about 2-3 drops are dropped into nose as a remedy for migraine.

Veterinary Medicine

1. A peg of stem is fixed in ground and cattle is tied to prevent from diseases and bad omens.
2. Stem bark dried or fresh is crushed in water and extract is given as antidote on snake bite.

Miscellaneous

1. Stumps of fresh wood/stem are fixed in crop fields of rice (*Oryza sativa* L.) Varai (*Panicum miliaceum* L.) to increase yield and protect the crop from devil's power.

Notes from Classics

Decoction of infusion of bark or root is applied on swellings. Bark is used as a virulent poison. Leaf powder is used as a fumigatory to rouse women from hysterical fits. Snuff of leaves is employed to relieve head-ache. Root is used as a antidote to snake-bite. Wood is used for cabinet works, combs and picture frames.

(Watt, 1889–1893)

Root extract is used as emetic. Paste of fresh root bark with water is applied on swellings. Leaves and root extract are used for tanning skins. Leaf powder is snuffed to get relief from head-ache and as a fumigatory in hysteria.

(Anonymous, 1948–1976)

Leaf powder is snuffed. Bark and leaves are used for tanning purposes. Wood is used for cabinet works, combs and picture frames.

(Ambasta, 1986)

Root bark is used on swellings and is said to be emetic, also used for inflammation of belly and gastric complaints. Stem bark is used in diarrhoea, hysteria, urino-genital complaints, and also as red dye. Leaves are used for fever. Bath of leaf decoction is given to mother after child-birth against rheumatism. Entire plant is useful as a remedy for bronchitis, cholera, epilepsy and syphilis.

(Jain, 1991)

Critical Notes

Latex is useful for cracks on lips, whereas leaf juice is valuable to treat migraine. Both uses and the ethnobelief noted are all unrecorded in past literature.

**Catunaregum spinosa* (Thunb.) Tiruv.

Family: Rubiaceae, Local Name: Gal

Ethnobotanical Notes

Medicine

1. Fruits are rubbed on stones and a cup of slurry is orally given for a fortnight to a person for madness.
2. Seeds paste is applied on neck as a remedy for tonsilitis.

Miscellaneous

1. Fruits are used for fish killing. Crushed fruits are spread in pond or lake to stupefy them.

Notes from Classics

Roots posses insecticidal and insect repelling properties. Bark is sedative, given to relieve pains due to bruises, bone-ache during fever, also used internally and externally as an anodyne in rheumatism. Fruit extract has insecticidal and insect repellent properties. Fruits are emetic, and used as a substitute for Ipecacuanha (*Cephaelis ipecacuahna*), in small doses it is expectorant and diaphoeretic. Fruits are used as fish poison. Wood is used for walking sticks, umbrella handles, ploughs, cotton-reels, bobbins, mathematical instruments and calico-printing blocks. Plant is used as goat fodder.

(Ambasta, 1986)

Roots are used on eye diseases. Stem bark is used on wounds and rheumatism. Fruits are used in amoebiasis, boils, dysentery, also used on stomach pains. Fruits and seeds are used as fish poison. Wood is used for making smoking pipes. Fruits are used as vegetable and also in drinks. Fruits are worn on waist of bride and groom to protect the stored grains. Fruits are symbolically sacred to God Shiva.

(Jain, 1991)

Critical Notes

Medicinal utility of fruits to treat madness is not reported in past literature.

**Cayratia auriculata* (Wall. ex Wight & Arn.) Gamble

Family: Vitaceae, Local Name: Ambatvel

Ethnobotanical Notes

Veterinary Medicine

1. Stem extract, about half cup only once, is given to children to control dysentery.
2. Stem pieces are crushed and extract, about 2 to 3 litres, is given to cattle on dysentery till it cures.

Notes from Classics

Stem pieces are cooked and eaten as vegetable. Leaves are edible in curry.

(Jain, 1991)

Critical Notes

Stem is being noted useful for the first time to treat dysentery for both, human beings as well as pet animals.

**Cayratia trifolia* (L.) Domin

Family: Vitaceae, Local Name: Teenpanya Kand

Ethnobotanical Notes

Medicine

1. Root paste is mixed with coconut oil and an applied for dislocations for three days.

Veterinary Medicine

1. Fresh twig of a plant is tied on horns of a bullock to cure swelling, injury and infection on shoulder or neck.

Notes from Classics

Roots are ground with black pepper and applied as poultice on boils. Decoction of root and leaf is recommended for fevers and perspirations. Poultice of leaves acts as rubefacient and is applied on to ulcers and yoke sores of bullocks.

(Anonymous, 1948–1976)

Ambasta (1986) gave uses similar to those in Anonymous (*loc.cit.*).

Roots are useful for anaemic person. Leaves are used to stop bleeding of injuries. Poultice of leaves is used on sores of neck of cattle. Leaves and fruits are edible.

(Jain, 1991)

Critical Notes

Root paste against dislocations in human beings and fresh twigs for veterinary purposes are not reported to date.

Celastrus paniculatus Willd.

Family: Celastraceae, Local Name: Kargangni

Ethnobotanical Notes

Medicine

1. Fruit oil is applied on joints in case of rheumatic pains.

Notes from Classics

The juice of leaves mixed with leaves of *Hydrocotyle asiatica* L. are powdered and considered a cooling application on inflammatory brain affection. Seeds are considered to be hot and dry, aphrodisiacal and stimulant, useful externally as well as internally as remedy on rheumatism, gout, paralysis and leprosy. Seed oil is diuretic and used in healing sinuses and fistulae, and also helpful in disorders of stomach.

(Watt, 1889–1893)

Bark is abortifacient. Seeds are said to be tonic and aphrodisiac. Seed oil is used as a nerve stimulant or brain tonic, also used for rheumatic pains.

(Ambasta, 1986)

Roots are used on sexual disorders. Stem bark is abortifacient, also applied on bone fracture, used on bronchitis and swollen veins. Fruits are used as an application for skin diseases and scabies and dysentery. Seeds are used on cold, cough, digestive complaint, fever, gout, leprosy, also used as hair tonic, stimulant, and also used as an luminant. Seed oil is applied on rheumatism, body-ache, eczema and stomach-ache. Plant is useful on anaemia, colic, sores and diarrhoea. Roots are used as antidote on snake-bite. Fruits are edible. Flowers are eaten as vegetable.

(Jain, 1991)

**Celosia argentea* L.

Family: Amaranthaceae, Local Name: Kirdu

Ethnobotanical Notes

Medicine

1. Root extract 10–20 ml daily at night for 7–8 days is given to control excessive menstruation.
2. Root pieces are tied around waist for easy delivery.
3. Half glass of root juice is given with hot water as antidote against snake-bite.

Veterinary Medicine

1. Root extract is mixed with curd in equal proportions. It is then diluted thoroughly and given to calf against dysentery.
2. Root pieces are tied on back of cows or buffaloes for easy delivery.

Miscellaneous

1. Leaves are cooked and eaten as vegetable.
2. Leaves are used as goat or cattle fodder.

Notes from Classics

Seeds are used on diarrhoea. Medicinal oil is also extracted from seeds. Plant is fed to buffaloes in times of scarcity.

(Watt, 1889–1893)

Seeds are considered efficacious in diarrhoea, also used in blood diseases, mouth sores and eye complaints for clearing vision. Plant is used as cattle fodder.

(Anonymous, 1948–1976)

Ambasta (1986) gave uses similar to those in Anonymous (*loc.cit.*).

Seeds are useful on diarrhoea, sores in mouth and also helpful in blood purification. Plant is used on blood dysentery. Tender shoots and leaves are cooked and eaten as vegetable.

(Jain, 1991)

Critical Notes

Roots are used variously against dysentery in animals, excessive menstruation, for easy delivery of animals and human beings and snake-bite etc. are not reported earlier.

****Chlorophytum tuberosum*** **(Roxb.) Baker**

Family: Liliaceae, Local Name: Musali

Ethnobotanical Notes

Medicine

1. Tubers are consumed raw when fresh as a general body tonic.

Miscellaneous

1. Leaves are cooked and eaten as vegetable.
2. Leaves are used as goat fodder.

Notes from Classics

Roots are edible.

(Anonymous, 1948–1976)

Roots are edible.

(Ambasta, 1986)

Tubers are expectorant and used in fever, leucoderma and also used as sex tonic. Leaves are cooked and eaten as vegetable.

(Jain, 1991)

Critical Notes

Roots are widely documented as useful tonic. Leaves are also valuable for both food and fodder. It is the experience of the present investigator that the root tubers are collected in the name of 'Safed-Musali', which is actually another species of the genus (*Chlorophytum borivilianum* Sant. & Fern.).

**Chrysopogon fulvus* (Spr.) Chiov.

Family: Graminae, Local Name: Kahandol

Ethnobotanical Notes

Medicine

1. Culms of this plant, bark of *Careya arborea* Roxb. and *Cassia fistula* L. are boiled together in water. This decoction about 10–15 ml orally is given on cough and chest pain.

Notes from Classics

Seeds are sometimes eaten by natives. Plant is used as excellent fodder.

(Watt, 1889–1893)

Plant is used as fodder, preferably cut before flowering. Plant also used as sand binder.

(Ambasta, 1986)

Critical Notes

Medicinal utility has not been reported of this taxon.

**Cissampelos pareira* L.

Family: Menispermaceae, Local Name: Tannivel,

Ethnobotanical Notes

Medicine

1. Roots are boiled and decoction is given in leucorrhoea.
2. Fresh leaf juice, about half a cup for three days, is useful in urine complaints.

3. Fresh leaves are crushed and few drops of juice is administered to baby only to treat greenish dysentery. Or, ring is made from stem and tied on right leg and left hand until it cures.
4. Twigs are crushed and extract is given to person of snake-bite to reduce poison.

Notes from Classics

Decoction of roots is administered on stomach pain, unhealthy sores and sinuses. Roots are considered as aromatic and bitter, prescribed in nephritic disorders, ulcers of kidney and bladder, also useful in jaundice and humural asthma. Dried root bark is mild tonic, used as diuretic in advance stages of acute and chronic cystitis, and catarrhal affection of bladder. It also exercises apparantly as astringent and sedative action on mucous membranes of genito-urinary organs. Leaves are applied on abcesses.

(Watt, 1889–1893)

Roots are used in dyspepsia, diarrhoea, dropsy, cough and cystitis. Leaves are externally used for sores and itch. Plant yields strong fibre.

(Anonymous, 1948–1976)

Ambasta (1986) mentioned uses similar to those in Anonymous (*loc.cit.*).

Roots are used as remedy on body-ache, catarrhal, infection of bladder, colics, cough, diarrhoea, dysentery, gastric complaints, head-ache, indigestion, leucoderma, leprosy, malaria, spleen enlargement, stomach ulcers and skin diseases. Roots and leaves are used against urine complaint. Stem is worn around the necks which gives relief from head-ache. Leaves are used on abcesses, antifertility, boils, cold, eye complaints, fever, heart burn, inflammation, jaundice, pimples, small-pox, stomach-ache and wounds. Leaves, flowers and roots together are useful on itching. Entire plant is useful for child-birth. Plant is useful on bronchitis, carbuncle, colic, diarrhoea, and epilepsy. Roots are used as antidote against snake-bite and scorpion sting. Rhizome is used in formation of local drinks.

(Jain, 1991)

Critical Notes

Roots for leucorrhoea, fresh leaves to treat urinary complaints, its juice as a remedy against greenish dysentery are new applications of this species. Plant species is exotic, native of South America.

Cissus quadrangularis L.

Family: Vitaceae, Local Name: Sandhejad

Ethnobotanical Notes

Medicine

1. Stem powder is made into paste and applied on fracture of bones in case of human beings.

Veterinary Medicine

1. Stem powder mixed with wheat flour 1 : 1 ratio. This mixture is fed to animals to cure fractures and dislocations.

Miscellaneous

1. It is believed that if the stem piece is broken in front of a drummer or one who beats dhol, his drum or dhol is torn out.

Notes from Classics

Stem juice is dropped in ears in ottorrhoea and into nose in epistaxis. It is also a remedy in scurvy and in irregular menstruation. Decoction of stem by boiling in lime water is considered as stomachic. Leaves and shoots when dried are powdered and given in bowel complaints.

(Watt, 1889–1893)

Root powder is used in the treatment of fractured bones. Fresh root powder is applied on burns and wounds. Stem paste is given on asthma. Powder of dry shoots is useful in digestive trouble. Plant juice is beneficial in scurvy, ottorrhoea and epitaxis. Tender shoots and young leaves are used in preparation of pappalans and curries. Stem and root yield strong fibres. The ash of plant is used as a substitute for baking powder. Powder obtained from shoots is applied on sores of saddle on horses and camel.

(Anonymous, 1948–1976)

Root powder is considered for septics and fracture. Stem juice is prescribed in scurvy. Roots and stem yield strong fibre. Young shoots are used in curries and also is preparation of papad.

(Ambasta, 1986)

Roots are used to cure wounds, against dysentery and indigestion. Stem powder is used on bone fractures, pains in joints and swellings. Leaves are useful on stomach-ache. Plant is used on wounds and fractures of cattle. Stem pieces are pickled and eaten on vegetable. Fruits are edible.

(Jain, 1991)

Critical Notes

It is usually planted in this region.

**Cissus repanda* Vahl

Family: Vitaceae, Local Name: Nandanvel

Ethnobotanical Notes

Veterinary Medicine

1. Stem piece is crushed at one end to make a brush. Cooked rice is mixed with salt, and applied on throat of cattle with the help of this brush to treat throat infection.

Miscellaneous

1. It is believed that if a twig of climber twined clockwise direction is kept in house. It keeps home happy and prosperous and if the climber is anti-clock wise causes quarrels and destructions in home.

Notes from Classics

Fibres are pulled from stem, used for rope making.

(Anonymous, 1948–1976)

Tubers are used in folliculosis. Bark is used on bone fracture. Leaves are used against sores, and useful for veterinary sores. Stem is used as edible vegetable.

(Jain, 1991)

Critical Notes

Use of stem on throat infection in case of animals is unreported in past. Ethnobobelief about the plant is interesting and also unrecorded earlier.

**Clematis triloba* Heyne ex Roth

Family: Ranunculaceae, Local Name: Mogra, Ranjai

Ethnobotanical Notes

Medicine

1. Leaf paste is applied on head to get relief from headache.

Veterinary Medicine

1. Twigs are crushed and slurry is applied on infected part of cattle.

Notes from Classics

Plant is used on leprosy, blood diseases and fever. Plant yields fibres which are used for various agricultural purposes.

(Watt, 1889–1893)

Plant is used as a remedy on leprosy, blood diseases and fever. Fibres are used for agricultural purposes.

(Anonymous, 1948–1976)

Uses recorded by Ambasta (1986) are similar to those of Anonymous (*loc.cit.*).

Plant is used in the treatment of boils and itches.

(Jain, 1991)

Critical Notes

To get relief from head-ache, leaves are not so reported in past.

**Clerodendrum multiflorum* (Burm.*f.*) O.Ktze.

Family: Verbenaceae, Local Name: Arni

Ethnobotanical Notes

Miscellaneous

1. Leaves are used as fodder especially for goats.

Notes from Classics

Root is bitter tonic and given in convalescence of measles. Juice of leaves is prescribed in neglected syphilitic complaints. Santals rub the plant over body in dropsy. Santals give the plant to cattle to cure diarrhoea, stomach swellings and worms.

(Watt, 1889–1893)

Roots are leaves are used on body-ache, headache and unconsciousness. Plant is said to be useful on colics, dysentery, dyspepsia, fever, post-natal, and stomach ache. Entire plant is specially used on mental diseases.

(Jain, 1991)

Critical Notes

Utility of leaves as fodder is not reported in past.

Clerodendrum serratum (L.) Moon.

Family: Verbenaceae, Local Name: Fangata, Bharangi

Ethnobotanical Notes

Medicine

1. Decoction of roots about 20 ml per day for a week is useful on joint pains.

Miscellaneous

1. Leaves and flowers are cooked and eaten as vegetable.

Notes from Classics

Roots constitute the *'Ghantu bharangi'*, exported from Vishakapatanam for medicinal purposes, used for febrile and catarrhal affections of natives. Roots boiled in water with ginger and coriander are given in nausea. Leaves boiled in oil are applied on ophthalmia. Seeds are boiled in butter milk, occasionally used in cases of dropsy. Leaves and flowers are eaten as vegetable.

(Watt, 1889–1893)

Roots are used on fever, dyspepsia and rheumatism. Decoction of root with coriander is given to reduce nausea. Leaves are useful on fevers. They are externally applied on cephalgia and opthalmia. Seeds are used in dropsy at some extent.

(Anonymous, 1948–1976)

Roots are used on appetite, body-ache, malaria, cough, fever and rheumatism. Roots are also used to mix in local drinks. Leaves are useful on eye complaints. Entire plant body is useful in anasarca, on cuts, dropsy, fevers and sores. Flowers are cooked and eaten as vegetable. Roots are used against snake-bite.

(Jain, 1991)

Clitoria ternatea Linn.

Family: Papilionaceae, Local Name: Gokarna

Ethnobotanical Notes

Miscellaneous

1. Plant is used for as fodder for goats.
2. Flowers are offered to God 'Mahadev'.

Notes from Classics

Root or root bark is a domestic medicine, administered to the children in croup and pulmonary affections. It acts as an emetic, nauseant, expetorant and laxative. When employed in larger doses to adults it acts as a demulcent, and diuretic and relieves some of the symptoms of gonnorrhoea, in irritation of bladder, as strangury, scalding of urine. Juice of leaves mixed with common salt, heated and is applied all around ear in ear-aches. Roots of blue flowered variety are used as an antidote in of snake-bite. Roots powder is useful in dropsy. Flower is held sacred to the Goddess 'Durga'.

(Watt, 1889–1893)

Roots are diuretic. Root bark is laxative. In Phillipines roots are used as poultice for swollen joints. Seeds mixed with ginger are laxative. Leaves are used as fodder for goats and sheep. Blue flowers yield dye, used as substitute for litmus paper. Seeds and root bark contain tannin.

(Anonymous, 1948–1976)

Roots are cathartic and diuretic. Flowers yield blue dye. Leaves are used as goat fodder. Seeds and root bark yield tannin.

(Ambasta, 1986)

Roots are abortifacient and purgative, used on goitre, stomach ache and leprosy. Seeds are laxative. Roots are useful on abdominal swellings of cattle. Plant is useful as antidote on snake-bite.

(Jain, 1991)

Critical Notes

The plant species is exotic, native of tropical America, grows wild, also introduced in gardens.

**Cocculus hirsutus* DC.

Family: Menispermaceae, Local Name: Nagan

Ethnobotanical Notes

Miscellaneous

1. Plant is used as fodder for cattle and goats.

Notes from Classics

Roots are bitter, alterative, laxative and demulcent, used with other ingredients on bilious, dyspepsia, rheumatism, and stomach ache of children. Leaf juice in water forms jelly, which is given as cooling medicine on gonorrhoea, and externally is useful on eczema, prurigo and impetigo.

(Anonymous, 1948–1976)

Ambasta (1986) described uses similar to those in Anonymous (*loc.cit.*).

Roots are useful in case of irritation of skin diseases, rheumatism and gout. Stem is used on stomach diseases. Leaves are used on eye diseases, conjunctivitis, cuts, eczema, cooling in fever, gonorhoea, leucorrhoea, nourishing mothers, sores, urine complaints, veneral diseases, wounds and as galactogogue. Stem and leaves are eaten as vegetable. Plant is useful on blood dysentery.

(Jain, 1991)

Critical Notes

The plant is useful as fodder. It is not reported in past literature.

Coccinia indica Wight & Arn.

Family: Cucurbitaceae, Local Name: Tondli

Ethnobotanical Notes

Miscellaneous

1. Fruits are eaten raw for sore mouth or infection in mouth.
2. Fruits are cooked and eaten as vegetable.

Notes from Classics

Roots, stems and leaves are efficacious in treatment of skin diseases, bronchial catarrah, bronchitis and diabetes. Fruits are cooked as vegetable when green; and eaten fresh when ripe.

(Anonymous, 1948–1976)

Raw fruits are used as vegetable, ripe ones are eaten. Tender shoots are used as pot-herb. Roots are used on head-ache and to stop vomiting. Leaves are useful in filirial swellings and loss of taste. Fruits are used on throat affection and diabetes. Entire plant is useful on convulsions, menorrhagia, scrofula, slow pulse, sores and syphilis. Fruits are eaten as vegetable.

(Jain, 1991)

**Coix gigantea* Koen. ex Roxb.

Family: Graminae, Local Name: Kachura

Ethnobotanical Notes

Medicine

1. Root stocks are cut and added in hot water. It is used for bath by a person suffering from joint pains. This is followed for 5-7 days.

2. A glass of extract of roots is administered orally to treat snake-bite.

Notes from Classics

Spherical grains are used by aborigines for ornamental purposes.

(Watt, 1889–1893)

Seeds are fed to poultry after separating from shells. Fruits are used for rosaries, bead curtains, ornamental trays, baskets and boxes.

(Anonymous, 1948–1976)

Ambasta (1986) gave uses similar to those in Anonymous (*loc.cit.*).

Critical Note

Administration of root extract to treat snake-bite is unreported.

Colocasia esculenta (Linn.) Schott.

Family: Araceae, Local Name: Tera

Ethnobotanical Notes

Miscellaneous

1. Leaves are cooked and eateh as vegetable.

Notes from Classics

Leaves and petiole juice is used as styptic or astringent. Tubers are peeled, sliced, cooked and taken with condiments. Taro flour can be prepared from raw or cooked tubers, can be used in soups, gruels, biscuits, cakes, bread pudding etc. Young leaves and stalks are cooked and used as spinach or sag.

(Anonymous, 1948–1976)

Ambasta (1986) has noted similar uses as those given by Anonymyous (*loc.cit.*).

Corm is used on cuts, wounds, injuries and on stings of bees. Plant is used to treat atrophy, bronchitis and cough. Corm is edible. Leaves are cooked and consumed as vegetable.

(Jain, 1991)

Commelina benghalensis L.

Family: Commelinaceae, Local Name: Kena

Ethnobotanical Notes

Miscellaneous

1. Leaves are cooked and eaten by natives as vegetable.
2. Entire plant is used as cattle fodder.

Notes from Classics

Plant is used as pot herb.

(Watt, 1889–1893)

Leaves are used for poulticing sores. Plant is used on leprosy. Leaves are cooked and eaten as vegetable. Leaves are also used as cattle fodder.

(Anonymous, 1948–1976)

Rhizomes are cooked and eaten. Herb is considered to be demulcent, emollient, laxatives, refrigerant, and also used on leprosy. Leaves are used as pot-herb in times of scarcity.

(Ambasta, 1986)

Roots are used on fever and liver complaints. Leaves are used on diarrhoea and irritation of bristles of catter pillars, also used on scorpion bite and wounds. Whole plant is considered to be refrigerant and is used against snake-bite and sores.

(Jain, 1991)

Critical Notes

The past literature and the present investigators observation indicate wide utility of this species for food value. It is a weed of cultivated lands.

***Commelina forsklaei* Vahl

Family: Commelinaceae, Local Name: Kenapet, Kula

Ethnobotanical Notes

Miscellaneous

1. Plant is used as cattle fodder.
2. Leaves are used as vegetable.

Critical Notes

This species is for the first time, being documented useful. It is a weed of cropped lands.

***Commelina hasskarlii* C.B.Cl.**

Family: Commelinaceae, Local Name: Kenpat

Ethnobotanical Notes

Miscellaneous

1. Entire plant is used as fodder for cattle.

Critical Notes

The plant species is for the first time being reported useful.

Commicarpus chinensis (L.) Heimerl.

Family: Nyctaginaceae

Ethnobotanical Notes

Miscellaneous

1. Entire plant is good fodder for cattle as well as goats.

Notes from Classics

Leaves are eaten as vegetable. Entire plant constitute as good fodder for goats.

(Jain, 1991)

***Commicarpus plumbagineus* (Cav.) Standl.**

Family: Nyctaginaceae

Ethnobotanical Notes

Miscellaneous

1. Plant is used as goat fodder.

Critical Notes

The species is for the time being reported useful.

Plate 1: A tribal man wearing 'Ghongada' or 'Irle' made up to of leaves of *Butea monosperma* (Lam.) Taub. and *Bambusa arundinacea* (Retz.) Willd., a stand of *Musa paradisiaca* L. and a plant of *Alocasia macrorrizha* (L.) Don in the foreground.

Plate 2: A Kokani tribal man at Hatavandi with 'Tondya', (A fish-trap) made up of strips of *Bambusa arundinacea* (Retz.) Willd. and rings of stems obtained from *Madhuca longifolia* (Koen.) MacBride var.latifolia (Roxb.) Chevalier.

Plate 3: *Ensete superbum* (Roxb.) Cheesm. in its natural habitat at Kamalichapada.

Plate 4: A tribal man with a stand made up of *Bambusa arundinacea* (Retz.) Willd. in hand used for egg laying at Jamumatha (Umarvihir).

Plate 5: A tribal Konkani family removing kernels from the seeds of *Madhuca longifolia* (Koen.) MacBride var.*latifolia* (Roxb.) Chevalier.

Plate 6: Tapping of 'Toddy' from *Phoenix sylvestris* Roxb.

Plate 7: A clump of *Colocasia esculenta* (Linn.) Schott. and a climber of *Lagenaria siceraria* (Mol.) Standl. growing near a house.

Plate 8: A tribal Kokani medicineman with different drugs in front at Payarpada (near Bathoda, Saptshringi, Dindori road).

Plate 9: Tribal Mahadevkoli farmer while breaking large soil lumps in field using a large 'Phali' (A large thick plate) made up from the wood of *Ougeinia oogeinensis* (Roxb.) Hochr.

Plate 10: Note the walking sticks of *Dendrocalamus strictus* (Roxb.) Nees at Devgaon Wadi to be marketed at Trimbakeshwar for the devotees of Lord Mahadev while circumambulating the Bramhagiri mountain.

Plate 11: A 'Thala' made of teak wood (*Tectona grandis* L.), used for storing flour while grinding food grains on hand-mill at Ambapani (Peth).

Plate 12: Note a tribal family and author (D.A. Patil) with fibres removed from the culms of *Bambusa arundinacea* (Retz.) Willd.

Plate 13: Illicit cutting of forest trees. *Anogeissus latifolia* (Roxb. ex DC.) Wall. ex Guill. & Perr. in view.

Plate 14: Note the bundles and bags containing flowers of *Curcuma pseudomontana* Grah. at Devgaon Wadi to be marketed at Trimbakeshwar as an offering to Lord Mahadeva.

Plate 15: Note the different varieties of jowar [*Sorghum bicolor* (L.) Moench. cultivated at Suryagad (Palasan Road).

Plate 16: A tribal Thakur boy holding in hand (i) 'Sagari', (A milk container), made up of *Bambusa arundinacea* (Retz.) Willd, a wooden doll carved out of the wood of *Cassia fistula* L. Note the pestle from the wood *Acacia catechu* Wight, a pillar of house prepared from entire trunk of teak tree (*Tectona grandis* L.) and the ladder in back side also from teak wood at Shrighat.

**Colebrookea oppositifolia* J.E.Sm.

Family: Labiatae, Local Name: Ran karvi

Ethnobotanical Notes

Medicine

1. Leaf juice is applied on injuries to cure early.

Miscellaneous

1. Stems are used for thatching walls of huts.
2. Dried stems are used as fuel.

Notes from Classics

Preparation of roots is used on epilepsy by Santals. Leaves are applied to wounds and bruises. Leaves are used as fodder for cattle. Wood is used for gunpowder charcoal.

(Watt, 1889–1893)

Santals use root preparation on epilepsy. Leaves are applied for wounds and bruises. Wood is used for making gunpowder and charcoal.

(Anonymous, 1948–1976)

Ambasta (1986) mentioned uses similar to those in Anonymous (*loc.cit.*).

(Ambasta, 1986)

Roots are used on hysteria and epilepsy. Stem is useful on cough. Leaves are used on burns, cuts, eye complaints, sores, wounds and to cure tooth-ache.

(Jain, 1991)

Critical Notes

The miscellaneous uses noted above are not reported earlier.

**Corchorus aestuans* L.

Family: Tiliaceae, Local Name: Chikan kharbat

Ethnobotanical Notes

Medicine

1. Root-stock part of this plant and that of *Cayratia trifolia*

(L.) Domin are crushed into paste which is applied on skeletal dislocations of human body.

Notes from Classics

A coarse fibre is sometime extracted from the plant.

(Watt, 1889–1893)

Plant yields fibre.

(Anonymous, 1948–1976)

Seeds of plant are stomachic.

(Ambasta, 1986)

Plant is used on fever. Leaves are said to be stomachic.

(Jain, 1991)

Critical Notes

Root stock is useful to treat dislocations. This is not reported earlier. The plant species is exotic, native of Tropical America.

**Corchorus olitorius* Linn.

Family: Tiliaceae, Local Name: Chunch

Ethnobotanical Notes

Medicine

1. Leaves are eaten as vegetable, which also helps to control dysentery in human being.

Miscellaneous

1. Leaves are used as fodder for goats.

Notes from Classics

Ashes of plant are mixed with honey and given daily in obstruction of abdominal viscera. Infusion of leaf is given internally as a drink in fever. Plant is used as a pot-herb. The stalks after removal of fibre are used for making gun powder charcoal, and also used in manufacture of baskets.

(Watt, 1889–1893)

Leaves are used as bitter tonic, diuretic and demulcent. Seeds are purgative. Leaves are eaten as pot-berb. Plant is a source of jute. It is used on / in fabrics, clothes, bandages etc. Jute cloth is used as wind screen, roofing, sand bags canvas, etc. Jute waste is used as filler for plastic.

(Anonymous, 1948–1976)

Leaves are used as substitute for spinach by Europeans. Infusion of leaves is good tonic, febrifuge, and also used as a demulcent in cystisis and dysuria.

(Ambasta, 1986)

Leaves are used on piles. Seeds are purgative. Leaves are edible.

(Jain, 1991)

Critical Notes

Uses of leaves as vegetable and also to control dysentery have remained still unnoticed. It is also found as a weed in cropped fields.

**Corchorus trilocularis* Linn.

Family: Tiliaceae, Local Name: Kaduchinchi

Ethnobotanical Notes

Miscellaneous

1. Plant is used as a goat and cattle fodder.

Notes from Classics

Seeds are bitter, administered in doses about 80 grams in fever and abdominal viscera obstructions. Seeds are specific in rheumatism.

(Watt, 1889–1893)

Plants yield fibre.

(Anonymous, 1948–1976)

Leaves are eaten as vegetable.

(Jain, 1991)

Critical Notes

Plant is useful as fodder. This application is not documented in past. It is found as weed of cultivated fields.

**Cordia dichotoma* Forst. *f.*

Family: Cordiaceae, Local Name: Bhokar

Ethnobotanical Notes

Medicine

1. Two cups of bark pieces are boiled in eight cup of water thoroughly till it becomes ¼. Decoction is given twice for four days on excessive menses.

Miscellaneous

1. Fruits are edible, these are either pickled or eaten when ripe.
2. Leaves in time of scarcity are fed to domestic animals.
3. Wood is used to make handles of agricultural tools.

Notes from Classics

Fruits are edible, and used as pickles, eaten as vegetable. Sticky juice is also used for book binding. Tender leaves are cooked and eaten as vegetable. Ripe fruits are boiled and decoction is given to patients of heart attack.

(Watt, 1889–1893)

Decoction of bark is used in dyspepsia and fevers. Fruits are astringent, anthelmintic, diuretic, demulcent, and expectorant used in diseases of chest, urinary passage and bronchitis. Kernels are used externally for ring worms. Fruits pulp is used as bird-lime. Wood is used for agricultural implements, constructions etc.

(Anonymous, 1948–1976)

Fruits are astringent, anthelmintic, diuretic, demulcent and expectorant, used in diseases of chest and urinary passage. Wood is used for boats, well-curbs, and agricultural implements, durable in contact with water.

(Ambasta, 1986)

Leaves are used on fever, headache, jaundice, colic, cough and ulcers. Fruits are used in the treatment of stomach-ache, urinal diseases, chest infection, lung diseases and constipation. Seeds are anthelmintic. Plant is useful on anasarca, cholera, dropsy, dysentery, dyspepsia, fever and utricaria. Fruits are edible.

(Jain, 1991)

Critical Note

Stem bark used against excessive menstruation is additional.

**Cordia gharaf* (Forssk.) Ehrenb. & Asch.

Family: Cordiaceae, Local Name: Gondhani

Ethnobotanical Notes

Medicine

1. Fruit juice is given as a remedy on spermatorrhoea and also reduces fevers.

Miscellaneous

1. Ripe fruits are edible.

Notes from Classics

Decoction of bark is astringent and used as a gargle. Fruits are eaten by poorer classes and is also pickled. Bark yields a coarse grey and white best fibre, used for rope making.

(Watt, 1889–1893)

Bark is chewed. Leaves are also chewed. Fruits are edible.

(Jain, 1991)

Critical Notes

To treat fever and spermatorrhoea, fruit juice are not yet noticed.

**Cordia mecleodii* Hook.*f*. & Thons.

Family: Cordiaceae, Local Name: Davisa

Ethnobotanical Notes

Medicine

1. Stem bark is boiled and half cup of decoction is given thrice a day for 3-5 days to anaemic persons.

Notes from Classics

Bark is used as a remedy for jaundice. Wood is used for making furniture, picture frames, and ornamental works, also used for fishing rods and bullock yokes.

(Watt, 1889–1893)

Wood is used for cart yokes, axe-handles, carving, turnery, picture frames and furniture.

(Anonymous, 1948–1976)

Uses recorded by Ambasta (1986) are similar as recorded by Watt (*loc.cit.*).

Critical Notes

To treat anaemic condition, stem bark is useful. This use does not find place in the past literature.

**Costus speciosus* (Koen.) J.E. Sm.

Family: Zingiberaceae, Local Name: Pewat

Ethnobotanical Notes

Miscellaneous

1. Tubers are crushed and spread in water bodies to stupefy fishes.

Notes from Classics

Roots are depurative, aphrodisiac, bitter, astringent, useful in catarrhal fevers, cough and skin diseases. Roots are tonic, anthelmentic, prescribed for pains in narrow by Santals. Roots of plant growing near water is used in 'Masalas' and inodorous. Tubers are cooked and made into syrups. Root stalk is a substitute for ginger.

(Watt, 1889–1893)

Roots are recommended as tonic. Rhizomes are eaten.

(Anonymous, 1948–1976)

Roots are used as tonic and anthelmintic. Rhizomes are cooked and eaten, accredited with purgative and tonic properties.

(Ambasta, 1986)

Rhizomes are anthelmintic, used against cold, constipation, ear complaints, eye complaints, head-ache, giddiness, leprosy, urine complaints, jaundice, indigestion and as galactogogue. Leaves are used on fever scabies, stomach diseases and wounds. Stem is used as maggoticide. Roots are used as veterinary medicine. Entire plant is used on anasarca, asthma, dropsy, cholera, bronchitis, and also useful on dog and jackal bite. Plant is said to be vermifuge, keeps snakes away. Roots are also as antidote on snake-bite.

(Jain, 1991)

Critical Notes

Employment of tubers as fish poison is not documented in past.

**Crinum defixum* Ker.-Gawl.

Family: Amaryllidaceae, Local Name: Kumbh

Ethnobotanical Notes

Medicine

1. Leaves are burnt and smoke is passed on body in case of inflamed swellings or spots on skin. It is treated for three days only.

Notes from Classics

Bulbs are boiled and eaten as shakh-bhaji.

(Watt, 1889–1893)

Bulb is nauseant, emetic, emollient and diaphoretic. It is used in the treatment of burns, whitlow and carbuncle.

(Anonymous, 1948–1976)

Bulbs are diaphoretic and emollient, are used for burns, whitlow and carbuncles.

(Ambasta, 1986)

Tubers are used on abdominal complaints and body ache. Leaves are used on ear-ache.

(Jain, 1991)

Critical Notes

Leaves to treat swelling is an additional application of this species.

**Crotolaria filipes* Bth.

Family: Papilionaceae, Local Name: Phatphati

Ethnobotanical Notes

Miscellaneous

1. Fodder: Plants are cherished by pet animals.

Critical Notes

Use of this species as fodder of domestic animals is noted for the first time.

Crotalaria pallida Ait.

Family: Papilionaceae, Local Name: Jungli tag, Velya karanj.

Ethnobotanical Notes

Medicine

1. Bark of this plant is burnt. This ash is called 'Misri'. It is smoked and this smoke is directed on body swelling.
2. Flowers and fruits are roasted, they are crushed and this paste is applied on inflammation till cure.

Notes from Classics

Root or stem piece is tied around wrist or neck of a person suffering from dropsy.

(Watt, 1889–1893)

Roots are used on swellings of joints. Leaves are used as vermifuge. Stem is used for cordages.

(Jain, 1991)

Critical Notes

The medicinal utilities such as bark for body swelling, flowers and fruits for inflammation are new records.

*Crotalaria retusa Linn.

Family: Papilionaceae, Local Name: Tiwara

Ethnobotanical Notes

Medicine

1. Branches are burnt and the ash called 'Misri' is applied on skin infections or scratching.

Miscellaneous

1. Flowers are cooked as vegetable by the tribals.

Notes from Classics

Plant yields fibre known as sun hemp for making cordage and canvas.

(Watt, 1889–1893)

Plant is used in the treatment of scabies and impetigo. Stem fibre is used for cordage and canvas. Plant is used as green manure.

(Anonymous, 1948–1976)

Plant yields fibre named sun hemp, used for cordage and canvas.

(Ambasta, 1986)

Critical Notes

Ash of stem to treat skin affection and flowers as edible are not noted earlier.

*Cryptolepis buchnani R.&S.

Family: Asclepiadaceae, Local Name: Bhui-kawali

Ethnobotanical Notes

Medicine

1. Root infusion about two spoons daily is used for seven days in the treatment of jaundice.
2. Leaf of *Tectona grandis* L.*f.* are rubbed on 'Dadari' (boils and pimples on neck) and latex of this plant is applied.

3. Roots of this plant, roots of *Kyllinga triceps* Rottb. and Ran wange (*Solanum virginianum* L.) are crushed together in babys urine mixture is boiled and cooled. Slurry obtained is applied on boils or tumours.

Miscellaneous

1. It is believed that a twig kept with us unknowingly to snake charmer, can fail his tricky game and tactics.

Notes from Classics

Plant preparation is combined with *Euphorbia microphylla* L. and given to mothers when flow of milk is deficient or fails. Preparation of plant is given to children on rickets.

(Watt, 1889–1893)

Fibres extracted from plants are used by hilly tribals.

(Anonymous, 1948–1976)

Plant yield fibres used by tribals for cordage and kind of cloth.

(Ambasta, 1986)

Roots are used against abdominal pain, fits, interim fever, rheumatism, stomach ache and helpful to increase lactation. Latex is applied on cuts and fissures of sole. Entire plant in useful for curing rickets in children, syphilis, veneral diseases, body-ache ascites cholera, dropsy, dysentery and sores. Plant is used as antidote on snake-bite.

(Jain, 1991)

Critical Notes

Roots are noted useful for jaundice and tumours. Latex is used as a remedy for boils and pimples. Ethnobelief noted above is also interesting. These have not been documented in past literature.

**Ctenolepis garcinii* (Burm.*f.*) C.B.Cl.

Family: Cucurbitaceae

Ethnobotanical Notes

Miscellaneous

1. Plant is used as fodder especially for goats.

Critical Notes

Utility of this climber as fodder for goats has still remained unnoticed.

Cucumis melo L. var.*agrestis* Naud.

Family: Cucurbitaceae, Local Name: Melu, Mekk.

Ethnobotanical Notes

Miscellaneous

1. Unripe or ripe fruits are edible.

Notes from Classics

Seeds are used in cooling medicine, also used in painful micturition and suppression of urine. Two drachms of seeds are rubbed into pulp with water, given along or with salt and 'Kanjika' as a powerful diuretic patient. Fruits are edible, also a substitute for cucumber. Seed oil is used for burning lamps.

(Watt, 1889–1893)

Seeds are edible and also as substitute for almond, pistachio. They are said to be diuretic, beneficial in chronic or acute eczema. Melons are eaten raw, jams, and syrups are also prepared from melons.

(Anonymous, 1948–1976)

Fruits are eaten as dessert. Seeds are edible, they diuretic, refregerant and nutritious. Seeds contain fixed oil. Seed pulp is used in chronic eczema.

(Ambasta, 1986)

Seed are used as cooling medicine. Fruits are edible.

(Jain, 1991)

**Cucumis prophetarum* Linn.

Family: Cucurbitaceae, Local Name: Chibbad

Ethnobotanical Notes

Medicine

1. Fruits along with 'morchud' (Copper sulphate) are crushed and mixture is applied on eczema.

2. Tribals infer that if a person is bitten by a snake the taste of fruits is felt sweet, otherwise it is always bitter.

Notes from Classics

Plant is emetic and purgative. It is highly toxic to animals if given orally.

(Watt, 1889–1893)

Critical Notes

Fruits on eczema and their tasting to verify the snake-bite are not yet documented elsewhere.

***Cucumis setosus* Cogn.**

Family: Cucurbitaceae, Local Name: Mekh

Ethnobotanical Notes

Miscellaneous

1. Unripe fruits are eaten raw.
2. Plant is used as cattle fodder.

Critical Notes

Both uses are new to the science of ethnobotany.

Curculigo orchioides Gaertn.

Family: Hypoxidaceae, Local Name: Kalacheda

Ethnobotanical Notes

Medicine

1. Root paste is applied on wounds to expel worms in case of human beings.

Notes from Classics

Tuberous roots are considered as alterative tonic, resortive, useful on piles, debility and impotence. Black 'Musali' (tuberous root) is used on asthma, piles, jaundice, diarrhoea, colics and gonorrhoea considered as demulcent, diuretic tonic and aphrodisiac.

(Watt, 1889–1893)

Two year old tuberous roots are sliced and dried in shade, and are said to be tonic, alterative, demulcent, diuretic and resortive. Tubrous roots are usually administered in combination with aromatics on piles, diarrhoea, jaundice and asthma, also used as poultice on itch or skin diseases.

(Anonymous, 1948–1976)

Ambasta (1986) mentioned uses similar to those in Anonymous (*loc.cit.*).

Roots are abortifacient, used on blindness, cuts, dysentery, ear complaint, epilepsy, white spots on eye ball, filaria, fever, impotency, itching, leucorrhoea, nose bleeding and pile bleeding. Rhizomes are used on asthma. Seeds are used as galactogogue. Roots are also used as tonic on unconsiousness of children, veneral complaints and wounds. Roots are used for cattle eyes in various complaints. Roots are used as antidote on snake-bite and scorpion sting. Roots are edible.

(Jain, 1991)

**Curcuma amada* Roxb.

Family: Zingiberaceae, Local Name: Ambe-halad

Ethnobotanical Notes

Medicine

1. Rhizomes are rubbed on stone and this slurry is applied on body of children suffering from fever. Or rhizomes are chewed and this extract is applied on body to reduce fever.

Notes from Classics

Tubers are regarded as cooling and used in prurigo, also employed as carminative and stomachic. Tuber paste mixed with spirit and white part of an egg is an application in chronic rheumatism and bruises, also applied over contusions and sprains. Roots are expectorant and astringent, useful in the treatment of diarrhoea and gleet. Tuber are used as condiment.

(Watt, 1889–1893)

Rhizome is used mostly in pickles and is considered to be stomachic and carminative. Rhizomes are applied on contusions and sprains.

(Anonymous, 1948–1976)

Rhizomes are carminative and stomachic, also used on contusions and sprains. Rhizomes are used in pickles.

(Ambasta, 1986)

Root stalk is eaten as vegetable. Roots are aromatic.

(Jain, 1991)

Critical Notes

Rhizome is being noticed useful for treating fever for the first time.

Curcuma domestica Valeton

Family: Zingiberaceae, Local Name: Halad

Ethnobotanical Notes

Veterinary Medicine

1. Rhizome powder is mixed with 'soot' collected from the house-roof above hearth. This mixture is rubbed on the tongue of cattle if the cattle are reluctant to eat fodder.

Notes from Classics

Rhizomes are stimulant, externally applied on pains and bruises, also administered internally in disorders of blood. Decoction of rhizome gives relief from catarrah and purulent opthalmia. Roots are paiched and powder is given in bronchitis and in catarrah. Roots are burnt and fumes are employed during hysteria and fits by witch-drivers. Rhizome paste is applied in case of vertigo. Turmeric is used in affection of liver and jaundice. Decoction of turmeric is used in purulent conjunctivitis. Fumes of burning turmeric in nostrils cause a copious mucous discharge and relieves congestion. Turmeric paste is applied in intermittent fever and dropsy. Paste of flowers is used against ringworms and other parasitic, skin diseases and scabies. Turmeric and alum 1:20 in proportion is blown into ear in chronic ottorrhoea. Turmeric forms one of the indispensable ingradients in curries, and also used for colouring confection. Smoke produced by sprinkling turmeric powder gives relief on scorpion-bite.

(Watt, 1889–1893)

Rhizome powder with lime forms good remedy for inflamed joints. Decoction of rhizome relieves pain of purulent opthalmia. Rhizome powder with lime forms paste, used as a remedy for inflamed joints. Juice of fresh rhizome is used as antiparasitic for skin diseases. Turmeric is used as a colouring material and as a condiment. Turmeric powder is prescribed as stomachic, tonic, blood purifier and antiperiodic alterative. Powder boiled in milk is useful for cough and cold. Turmeric is used as a colouring matter in pharmacy, confectionary and food industry. Turmeric paper is a official reagent for testing alkalinity. A diluted tincture of turmeric is suitable for use of fluorescence indicator. Turmeric powder is auspicious powder in all redigious observations in Hindu households.

(Anonymous, 1948–1976)

Rhizome are stimulant, stomachic, depurative and as tonic. Rhizomes are used as condiment and as colouring agent. Curcuma paper is used as indicator of chemicals. Turmeric is used in composition for sprains and bruises.

(Ambasta, 1986)

Roots are used in leprosy, migraine, body pain, pimples and feckles on face, body swelling, wounds and sacred in religious rites. Leaves are used against cold, fever and pneumonia. Rhizomes are said to be anthelmintic. Flowers are used in case of antifertility. Plant is useful in cases of atrophy, blindness, bronchitis, carbuncle, cholera, dyspepsia, indigestion, rheumatism, scabies, sore, syphilis, spleen complaints and eye diseases.

(Jain, 1991)

Critical Notes

The rhizome powder is being reported for the first time useful to treat animal diseases. It is usually not cultivated in this region but its rhizomes are sold in market.

Curcuma pseudomontana Graham.

Family: Zingiberaceae, Local Name: Shilandh

Ethnobotanical Notes

Medicine

1. Extract of rhizomes, about 20 ml at night, is administered orally to increase lactation and better digestion in ladies.

Miscellaneous

1. Inflorescences are offered to God 'Mahadeva' of Trimbakeshwar, sold at cost Rs. 3/- each.
2. Rhizomes are cooked and eaten in times of scarcity by natives.

Notes from Classics

Tubers are boiled and eaten by natives in times of scarcity.

(Watt, 1889–1893)

Rhizomes yield a form of arrow root.

(Ambasta, 1986)

Rhizomes are eaten as vegetable.

(Jain, 1991)

Critical Notes

Rhizomes are used to increase lactation. Inflorescences are offered to Lord 'Mahadeva'. Both uses are not documented in past. It is to be noted that removal of rhizomes and large collections of inflorescence appeared to have reduced the populations of this species considerably in this area. There is a dire necessity to conserve this species before it meets a sorry state of affair.

**Cuscuta reflexa* Roxb.

Family: Cuscutaceae, Local Name: Amarvel

Ethnobotanical Notes

Veterinary Medicine

1. Stem is crushed and 3 to 4 glasses of extract it is given to the animals per day against dysentery till cure.

2. Stem powder added with wheat flour is passed over body from tail to head of a cows or she-buffaloes to increase lactation.

Notes from Classics

Stem is useful in bilious disorders. Cold infusion is given as depurative, also used externally on itch and internally on fever. Seeds are carminative, boiled and warm seeds are applied over stomach as an anodyne. Plants sometimes are used as dye. Plant is used as goat fodder.

(Watt, 1889–1893)

Seeds are said to be diaphoretic, demulcent and tonic. Decoction of plant is used as wash on sores. Plant is purgative and is used internally on protracted fevers,retention of winds, and induration of liver. It is applied externally on itch.

(Anonymous, 1948–1976)

Seeds are regarded as diaphoretic, dumulcent and tonic. Plant is considered to be purgative, used in flatulence and liver complaints. Entire plant paste is externally used on itch.

(Ambasta, 1986)

Stem is used on indigestion, jaundice, and liver complaints. Seeds are used against stomach-ache. Entire plant is used medicinally against bilious body-ache, bone fracture, burns, cuts, cold, cough, dipthera, displaced knee-cap, eye diseases, fall of hair, headache, itching, oedema, swelling, legs, testicles, body wounds, body-ache, and to kill lices. Plant is useful as veterinary medicine.

(Jain, 1991)

Critical Notes

Stem is useful to treat dysentery in case of mankind and increase lactation of milking animals. These medicinal application are hitherto unreported.

Cymbopogon martinii (Roxb.) Wats.

Family: Graminae, Local Name: Rosha

Ethnobotanical Notes

Medicine

1. Oil distilled from grass is used on joints, pains and rheumatism.
2. Plant is used as fodder for pet animals.

Noteş from Classics

Oil obtained from plant is used in chronic rheumatism and neuralgia. Natives consider it as cooling and astringent and hence useful in skin diseases, and bilious affections. A spirit of plant is used in indigestion and fever. Plant is favourite fodder for cattle.

(Watt, 1889–1893)

Stem, leaves and flowers yield 'Palmrosa' oil which is used in adulteration of attar of roses, also used in several perfumes, cosmetics and soaps. Alongwith sandal wood oil it is used in ointment for warding of mosquitoes, used in medicine on lumbago, stiff joints and skin diseases. It also cures baldness, taken in small doses internally as remedy for bilious complaints.

(Anonymous, 1948–1976)

Plant is a source of Palmrosa oil, used in soaps, cosmetics, also used on lumbago and stiff joins. Extensively used in India as adulterant of attar of roses.

(Ambasta, 1986)

Leaves are used on rheumatism. Oil extracted from plant is used on bilious, rheumatism, and skin diseases.

(Jain, 1991)

Cynodon dactylon (L.) Pers.

Family: Graminae, Local Name: Hariyali

Ethnobotanical Notes

Medicine

1. Leaves are crushed and juice is applied on bleeding injuries.

Miscellaneous

1. Plant is used as fodder for cattle.
2. Culms are offered to lord 'Ganesha'.

Notes from Classics

Roots are crushed and mixed with curd are used in cases of chronic gleet. Decoction of roots is used in secondary syphilis, used as diuretic. Fresh juice of leaves is astringent and used in epitaxis. Leaf juice is also applied on wounds, and useful in chronic diarrhoea and dysentery. Juice of entire grass is used in catarrhal opthalmia. Leaf juice with milk is given on irritation of urinary organs. A cooling drink can be made from roots. Plant is good fodder for horses and cows. Plant is sacred to Lord Ganesha.

(Watt, 1889–1893)

Rhizomes are used on genito-urinary disorders. Rhizomes and roots are used as adulterant and substitute for *Agropyron repens*. Beauv. (Couch grass). Decoction of plant is diuretic, considered useful in dropsy and anasarca. Plant juice is astringent, applied on bleeding cuts and wounds.

(Anonymous, 1948–1976)

Rhizomes are used in genito-urinary troubles. It is important as pasture and lawn grass. Plant decoction is diuretic, used in anasarca. Plant in good for soil binder.

(Ambasta, 1986)

Roots are cooling, used in dysentery, hysteria, opthalmia, and bleeding piles. Stem is used as tonic. Leaves are used on dropsy, urine complaints, bleeding piles, and control vomiting. Plant is used on carbuncle, cramps, head-ache, menstrual complaints, ulcers in stomach and wounds. Plant is symbolically sacred. The roots are abode of Bramha, stem abode of Vishnu and leaf tip abode of Shiva.

(Jain, 1991)

Cyperus pangorei Rottb.

Family: Cyperaceae, Local Name: Gundon

Ethnobotanical Notes

Miscellaneous

1. Entire plant is used for thatching roofs of huts.

2. On the tenth day of funeral two persons worship for his soul. A stem piece is given to both of them which is later interchanged, broken at center and thrown in river, or lake, or well etc. This trend is followed by Hindu Kokani people.

Notes from Classics

Plant is eaten by buffaloes as fodder. Plant is used for mat making.

(Watt, 1889–1893)

Plant is used as antidote on snake-bite.

(Jain, 1991)

Critical Notes

It is used for completing a ritual after death of a Kokani person.

**Cyperus rotundus* L.

Family: Cyperaceae, Local Name: Lawahali

Ethnobotanical Notes

Medicine

1. Tuber paste mixed with honey is consumed for yellow urination. It is followed for 3 to 4 days.

Miscellaneous

1. Entire plant is used for thatching huts.

Notes from Classics

Roots are considered to be diaphoretic, astringent and vermifuge, recommended to cure stomach disorders, irritation of bowels, fever, diarrhoea, and dysentery. Bulbous roots with ginger and honey is given in cases of dysentery. In Konkan paste of tubers is applied on breast of ladies to increase flow of milk. Tuber paste is applied on scorpion-bite and dried tubers are applied on spreading ulcers. 'Theofrastus' mentioned it as favourite food of horses. Rhizomes yield oil which is used in perfumes. Dye is used in certain dye preparations to import a perfume to a fabric.

(Watt, 1889–1893)

Tubers are aromatic, used in medicine, perfumery and Agarbatties. Tubers are given on disorders of stomach and irritation of bowels.

(Anonymous, 1948–1976)

Ambasta (1986) reported uses similar to those in Anonymous (*loc.cit.*).

Roots are astringent, used against bowel complaints, diuretic, also used for jaundice, sores, wounds, stomach diseases and mastics. Bulbs are used on delirium, dysentery and urine complaints. Plant is used on heat-stroke. Roots are used as antidote on snake bite.

(Jain, 1991)

Critical Notes

Application of root tubers to treat yellow urination is new and noteworthy.

**Dalbergia lanceolaria* L.f.

Family: Papilionaceae, Local Name: Dandus

Ethnobotanical Notes

Medicine

1. Stem bark of 'Asand' [*Bridelia airy-shawii* P.T.Li.] and of this plant are crushed in water. Few drops of this filtrate are poured in ears to get relief from ear-ache.
2. Paste of stem bark is applied on scorpion sting.

Notes from Classics

Milky latex exudates from root is applied on ulcers. Infusion of bark is given internally in dyspepsia. Leaves are rubbed over the body as a treatment for leprosy and other cutaneous diseases.

(Watt, 1889–1893)

Decoction of bark is used in dyspepsia. Seed oil is applied on rheumatism. Wood is used for making agricultural implements, also suitable for carving, boarding, rafters, scantlings, packing cases and for general construction purposes.

(Anonymous, 1948–1976)

Decoction of bark is used in dyspepsia. Wood is used for agricultural implements, also suitable for carving, boarding, rafters, packing cases and construction.

(Ambasta, 1986)

Stem bark is useful on fever and tongue infection. Twigs are used as tooth-brush.

(Jain, 1991)

Critical Notes

Stem bark is employed against ear-ache and scorpion sting. Both uses do not find place in classical literature.

Dalbergia sissoo. Roxb.

Family: Papilionaceae, Local Name: Shisam

Ethnobotanical Notes

Miscellaneous

1. Wood is used for making furnishes.
2. A 'Musal' (Pounding pestle) is made for pounding grains.

Notes from Classics

Roots are considered astringent. Bark pills with aromatics like ginger are used as a remedy for cholera. The raspings of wood is regarded as alterative, useful in leprosy, boils, eruptions, and the allay vomiting and special diseases. Decoction of leaves is useful in acute stages of gonorrhoea. Wood is durable, and is used for boat making, carts, carriages, agricultural implements, furniture, and house constructions.

(Watt, 1889–1893)

Bark paste is applied on pimples. Decoction of leaves is used on gonorrhoea. Leaf mucilage mixed with sweet oil is used as an application for excoriations. Heart wood yields viscous fixed oil which on cooling becomes solid like vaseline. Oil is used as lubricant for heavy machinery. Wood is used for high class furniture, railway sleepers, musical instruments, electric casing, charpai legs, tool handles, tobacco pipes, laminated skis, plywood logs, and ornamental carvings, also used for making charcoal.

(Anonymous, 1948–1976)

Leaves are used as cattle fodder. Wood is useful for furniture, railway sleepers, musical instrument, charpais, carvings etc., also suitable for charcoal making.

(Ambasta, 1986)

Stem is used in the treatment of head-ache and leprosy. Leaves are said to anthelmintic, used on dysentery, eruptions, scurf, skin diseases and gonorrhoea. Seeds are used as pendant in ornaments.

(Jain, 1991)

**Dalbergia volubilis* Roxb.

Family: Papilionaceae, Local Name: Yeloshi

Ethnobotanical Notes

Medicine

1. Petiole of curved leaf is made into a ring. It is placed in the ear hole which gives relief from migraine.

Notes from Classics

Roots with seeds of *Carum carvi* are given internally as a remedy on gonorrhoea. Leaves are used as fodder.

(Watt, 1889–1893)

Root juice mixed with sugar and cumin is given on gonorrhoea. Leaf juice is applied in apthae, and used to gargle sore throat. Leaves are used as cattle fodder.

(Anonymous, 1948–1976)

Ambasta (1986) mentioned uses similar to those in Anonymous (*loc.cit.*).

Leaves and roots in combination are used as to increase sex weakness.

(Jain, 1991)

Critical Notes

Petiole is thought beneficial to treat migraine.

***Dendrobium barbatulum* Lindl.**

Family: Orchidaceae, Local Name: Zadya-losan

Ethnobotanical Notes

Medicine

1. Tubers are crushed in water and the juice obtained is given to improve appetite. About 20 ml of it is advised at morning for 3-4 days.

Critical Notes

The plant species itself and especially tubers are being reported useful for the first time.

Dendrocalamus strictus (Roxb.) Nees

Family: Graminae, Local Name: Bamboo

Ethnobotanical Notes

Veterinary Medicine

1. Decoction of leaves is given to horses for throat infection.

Miscellaneous

1. Stem is used as walking stick while Circumambulation to Bramhagiri mountain at Trimbakeshwar.
2. From fibre 'Nahar' are prepared which are used for tying huts and roofs of huts.
3. Food grain containers (kothi) are also prepared from bamboo strips.
4. Bamboo stand is made for hens used for laying eggs.
5. Stems are used for hedges.
6. Young shoots (buds) are boiled and cooked as vegetable.

Notes from Classics

The silicious matter found near joints of stem is used as cooling tonic and astringent medicine. Decoction of internode is used as medicine to procure abortion. Seeds are eaten by natives in time of scarcity. Decoction of leaves is given to a lady after delivery to put uterus in order. Two ounces of leaf juice is taken frequently to cause abortion. Leaves are fed to animals during parturition from a supposition that they cause a more rapid expulsion of placenta. Leaves are fed to horses suffering from cough. Bamboo is used as a substitute for timber, to make rafters, buttons, walking sticks, fencing, mats, slates, umbrella handles, chairs, sofas and several minor uses.

(Watt, 1889–1893)

Make bamboo is employed as a raw material for paper and mathematical instruments.

(Anonymous, 1948–1976)

Stem is employed for rafters, buttons, scaffoldings, mats, baskets, sticks, furniture, tentpoles, water pipes, fishing rods, musical instruments, mats, for country boats, used as a bouyage of heavy timber, in rafter, also used in paper manufacture. Leaves are used as cattle fodder.

(Ambasta, 1986)

Roots are used in the treatment of fever and antifertility. Bark is used on injuries. Shoots are considered as tonic and eaten cooked, also used for basketery, cordage and other miscellaneous article. Leaves are used for cough. Leaves are used as veterinary medicine. Seeds are eaten as grains.

(Jain, 1991)

Critical Notes

Edibility of young shoots is already proved. The various local uses although not documented so, are obviously used as per the local needs of the tribal people. It is but apparent that this species should be always given place in afforestation programmes.

**Dendropthoe falcata* (L.*f*.). Ettingshausen

Family: Loranthaceae, Local Name: Bandgool

Ethnobotanical Notes

Medicine

1. Stem pieces of a plant growing on *Tamarindus indicus* L. are selected and red lead is sprinkled on them. Stem pieces with black thread are tied on loin of a pregnant women to prevent abortion.

Miscellaneous

1. Fruits are eaten by tribal children.

Notes from Classics

Bark is astringent and narcotic, useful in the treatment of menstrual complaints, wounds and also a remedy for consumption, asthma and mania. It is reported to be substitute for betel nut.

(Anonymous, 1948–1976)

Ambasta (1986) reported uses similar to those in Anonymous (*loc.cit.*).

Roots are used to treat skin diseases. Stem or entire plant is said to be abortifacient, used on antifertility, head-ache, paralysis, wounds, impotency and as a brain tonic.

(Jain, 1991)

Critical Notes

To fasten stems pieces around loin to prevent abortion is not noted earlier.

Desmodium dichotomum (Willd.) DC.

Family: Papilionaceae, Local Name: Asud.

Ethnobotanical Notes

Miscellaneous

1. A fresh twig is wiped on a body for seven times, it is believed that a person is freed from devil's or ghost's power.

Notes from Classics

Foliage of plant is eaten by cattle.

(Watt, 1889–1893)

Anonymous (1948–1976), Ambasta (1986) and Jain (1991) mentioned use similar to Watt (*loc.cit.*).

Critical Notes

It is an ethno–belief that the plant frees a person from power of ghost or devil.

Desmodium gangetium (L.) DC.

Family: Papilionaceae, Local Name: Asud

Ethnobotanical Notes

Miscellaneous

1. Leaves are passed seven times over the body of a person suffering from fever.
2. A twig is passed over the seven times to get rid off ghost's power.

Notes from Classics

Root is one of the constituent of 'Dashmula' of Hindu Materia medica. Plant is regarded as a febrifuge and anti-cattarrhal.

(Watt, 1889–1893)

Roots are used as a febrifuge, bitter tonic, expectorant, alterative and diuretic. Stem fibre is used in paper making. Plant is used as green manure.

(Anonymous, 1948–1976)

Roots are used as a febrifuge, expectorant and diuretic. Plant is suitable for paper making.

(Ambasta, 1986)

Roots are used in the treatment of cough, asthma, fever, dysentery and vomiting. Entire plant is used as a remedy on bronchitis, also used on eye infection and as a tonic. Roots are used as antidote to snake venom. Stem fibre is used for paper making.

(Jain, 1991)

Critical Notes

This species of the genus has more or less similar uses as *D.dichotomum* in the district. Both are known as 'Asud'.

**Dichanthium pertusum* (L.) Clayton

Family: Graminae, Local Name: Kathur

Ethnobotanical Notes

Miscellaneous

1. Culms are used for thatching roofs of huts.

2. It is also used to tie fodder or fuel wood bundles.

Critical Notes

It is employed for thatching purpose and to fasten wood bundles. The species is reported useful for the first time.

**Dichrostachys cinerea* (L.) Wight & Arn.

Family: Mimosaceae, Local Name: Yeltur

Ethnobotanical Notes

Medicine

1. Roots are rubbed on stone and slurry obtained is applied on a plummet.

Miscellaneous

1. Wood is useful for making handles of axe, agricultural implements and pegs to tie to it the pet animals.
2. Leaves are fed to goats.

Notes from Classics

Roots are astringent and used in rheumatism, urinary calculi and renal troubles. Tender shoots are bruised and applied onto eyes in case of opthalmia. Leaves are used as cattle fodder. Heart wood is used as walking sticks and for tent pegs.

(Anonymous, 1948–1976)

Critical Notes

Roots are useful to treat plummet, an application hitherto unreported.

Digera muricata (L.) Mart.

Family: Amaranthaceae, Local Name: Kurdu

Ethnobotanical Notes

Miscellaneous

1. Leaves are eaten as vegetable.
2. Leaves are used as fodder for cattle.

Notes from Classics

Flowers and fruits are prescribed for urinary discharges. Tender shoots are eaten as pot-herb. Plant is relished by cattle.

(Ambasta, 1986).

Leaves are used on boils. Seeds and flowers are used on urine complaints.

(Jain, 1991)

Critical Notes

Food value of foliage of this species should be scrutinized on modern scientific lines. Plant species is exotic, native of North Africa.

**Dillenia pentagyna* Roxb.

Family: Dilleniaceae, Local Name: Karwal

Ethnobotanical Notes

Medicine

1. Stem bark extract prepared in water, about 15-20 ml, is administered in case of jaundice for a week. It is also used for urinal complaints.

Miscellaneous

1. Leaves are used as platters for dinner in social functions and ceremonies.

Notes from Classics

Bark is used for making cordage. Flower buds and fruits are eaten when green. Leaves are used for polishing ivory and horns. Wood is used in construction of ship, rice mills, and also for charcoal.

(Watt, 1889–1893)

Buds and fruits are eaten raw or cooked. Leaves are used as green manure. Wood is used for house-posts, raftares, plankings, almirahs, match making and Paneling.

(Anonymous, 1948–1976)

Ambasta (1986) gave uses similar to those in Anonymous (*loc.cit.*).

(Ambasta, 1986).

Bark is useful at the time of child-birth. Extract of stem bark and roots is used on rheumatism. Fruits are considered as cooling and used against diarrhoea and dysentery. Fruits are eaten as vegetable. Bringing of fruits in a village before April is a taboo.

(Jain, 1991)

Critical Notes

Stem bark is useful to treat jaundice. It is also used for urinal complaints this uses are hitherto not reported. Leaves are employed for making dinning plates. This is also not noted, although they are noted for some similar other uses.

**Dioscorea bulbifera* L.

Family: Dioscoreaceae, Local Name: Kadu kand

Ethnobotanical Notes

Medicine

1. Tubers are boiled, peeled and consumed. Fruits are eaten to reduce acidity.

Notes from Classics

Powder of tubers is applied on ulcers. Roots are employed for washing woolen, silk, cotton and clothes. Tubers are used as bait for fish. Stem bulbils and tubers are cooked and eaten a s vegetable. Leaves are used as cattle fodder.

(Watt, 1889–1893)

Dried pounded tubers are used as an application on ulcers and piles, also used as remedy on dysentery and syphilis. Bulbils of wild species are used as application for sores. Tubers are mostly used as famine food. Tubers are useful in washing of wooden clothes. Tuber pieces are used as fish bait (prey).

(Anonymous, 1948–1976)

Tubers are dried and pounded and applied on ulcers also used in the treatment of dysentery and piles.

(Ambasta, 1986)

Tubers are prescribed on abdomen pain, dysentery piles and jaundice. Tubers are eaten as vegetable.

(Jain, 1991)

Critical Notes

Consumption of fruits to reduce acidity is unknown.

**Dioscorea hispida* Dennst.

Family: Dioscoriaceae, Local Name: Wachkand

Ethnobotanical Notes

Veterinary Medicine

1. Tubers are crushed and mixed with the fruits of *Carum carvi* Linn., *Piper nigrum* Linn. and *Myristica fragrans* Houtt. These are again crushed thoroughly. The filtrate is given to cattle suffering from 'phakya' disease on thigh.

Miscellaneous

1. Tubers are eaten after boiling and peeling. They are also added in fermentation of alcohol to increase percentage of alcohol.

Notes from Classics

Tubers are acrid and poisonous, used as food after proper washing, eaten only in times of food scarcity.

(Ambasta, 1986)

Entire plant is administered on boils, sores and whitlow, also used as a remedy on rabid dog and jackal bite. Tubers are eaten after several boiling.

(Jain, 1991)

Critical Notes

Tubers are found useful as ethno-veterinary medicine. They are also used for alcohol preparation. Both uses are not documented in past.

**Dioscorea oppositifolia* L.

Family: Dioscoreaceae, Local Name: Chaikan

Ethnobotanical Notes

Medicine

1. Tubers are chewed orally to control dysentery.

Miscellaneous

1. Leaves and bulbils are cooked and eaten as vegetable.
2. Young inflorescence are also consumed raw.

Notes from Classics

Root powder is applied on swellings. Roots are also useful as antidote on snake bite and scorpion-sting. Tubers and roots are eaten at the time of scarcity. Leaves are used as cattle fodder.

(Watt, 1889–1893)

Tubers after heating are used to reduce swellings. Tubers are edible.

(Anonymous, 1948–1976)

Tubers are ground and applied on swellings.

(Ambasta, 1986)

Tubers are used in the treatment of body-ache, constipation and swellings, also used as tonic and appetizer. Plant is useful on dysentery, hydrocele and sores.

(Jain, 1991)

Critical Notes

Edibility of inflorescence is unreported. This practice appears to have reduced the plant in the district. It reduces the chances of pollinations and ultimately fruits-set in the species. Tribals should be made aware of the fact. A stringent action is needed to conserve the taxon.

**Dioscorea pentaphylla* Linn.

Family: Dioscoreaceae, Local Name: Tamboli, Ulshi

Ethnobotanical Notes

Miscellaneous

1. Leaves are cooked and eaten as vegetable.
2. Young green fruits used as vegetable.

Notes from Classics

Tubers and flowers are cooked and eaten as vegetable. Leaves are eaten in times of scarcity.

(Watt, 1889–1893)

Tubers are used on swellings and also used as tonic. Flowers and leaves are eaten as vegetable in times of scarcity.

(Anonymous, 1948–1976)

Tubers are considered as tonic and used on swellings. Tubers are edible. Leaves and flowers are consumed as vegetable in times of scarcity.

(Ambasta, 1986)

Tubers are used against asthma, bile, cough and fracture of bones. Entire plant is used in the treatment of abdomen pain after labour, anasarca, dropsy and rheumatism.

(Jain, 1991)

Critical Notes

Use of green fruits as vegetable is unreported.

**Dioscorea wallichii* Hook.f.

Family: Dioscoreaceae, Local Name: Chaichamur

Ethnobotanical Notes

Miscellaneous

1. Inflorescence of female flowers are used as vegetable.
2. Young fruits are also eaten as vegetable.

Critical Notes

Uses of flowers and young fruits as vegetable are additional. However, this has hampered the populations of the species. There is a dire necessity to check its depletion.

Diospyros melanoxylon Roxb.

Family: Ebenaceae, Local Name: Temburni

Ethnobotanical Notes

Medicine

1. Stem pieces are burnt in hearth and this smoke is passed over body of a pregnant women for easy delivery and good health after delivery.

Veterinary Medicine

1. Stem pieces are collected and red lead (Shendur) is sprinkled on them. Hymns are murmerred and with the help of fibre or thread stem pieces are tied around neck of cattle as a remedy for injury and septics.
2. Pegs made from stem pieces are fixed in ground or cattle shed and milching animal are fastened to it. It is believed that it helps for continuous milk flow.

Miscellaneous

1. Ripe fruits are eaten.
2. Leaves are used for wrapping bidis.

Notes from Classics

Decoction of stem bark is used on diarrhoea, dyspepsia, as a tonic and also as lotion for eyes. Bark powder mixed with pepper is given on dysentery. Ripe fruits are edible. Leaves are used for rolling cigarettes and bidis. Wood is used in construction of houses.

(Watt, 1889–1893)

Leaves are diuretic, carminative and astringent. Decoction of leaves is useful in diarrhoea and dyspepsia. Dilute extract of leaves is used as lotion for eyes. Fruits are edible, carminative and astringent. Leaves are used for rolling bidies. Sap wood is used for rafters, shafts, poles of carts, plumber tools and agricultural implements. Black wood is used for making butts of billiard cues, carvel walking sticks, brush backs, snuff boxes, combs, toys and picture frames etc. Wood is used as fuel.

(Anonymous, 1948–1976)

Decoction of bark is used in diarrhoea and dyspepsia. Leaves are diuretic, laxative, carminative and styptic. Flowers are used on skin diseases and urinary complaints. Leaves are used for wrapping bidies. Fruits are edible.

(Ambasta, 1986)

Flowers are used for skin diseases. Fruits are used as germicide, urine diseases and on cracks of sole fruits. Wood is used symbiolically for idols. Excess fruiting on plant is indicates of post famine periods.

(Jain, 1991)

Critical Notes

Stem is thought beneficial during pre-and post-delivery period. It is helpful for cattle to treat their injury or septic. It is an ethno-belief that stems are helpful to maintain continuous flow of milk in case of milking animals.

**Diplocyclos palmatus* (L.) Jeffery

Family: Cucurbitaceae, Local Name: Shivlingi

Ethnobotanical Notes

Medicine

1. Seeds of this plant, one spoon lemon juice and cow-ghee are given to a woman on first day of menstruation for conception. It is administered twice a day for 4-5 days after bath. Or tablets of this mixture adding jaggery and prepared. One tablet twice a day is advised.
2. Paste of entire plant is applied on septics of human injuries as well as useful for cattle.
3. Seeds are kept in 'Tait' (amulet) and tied around neck or arm to protect from devil's or demon power, preferably on day of 'Mahashivratri and 'Hindu festival'.

Notes from Classics

Fruits are considered having tonic properties. Leaves are boiled and eaten as vegetable.

(Watt, 1889–1893)

Leaves are sometimes cooked and eaten.

(Anonymous, 1948–1976)

Herb is used as bitter tonic and febrifuge. Paste of leaves is used for boils. Cooked young fruits are consumed as sajoor, a soup.

(Ambasta, 1986)

Leaves are used on eye infections. Seeds promote fertility in women. Entire plant is used as a remedy on ague, also used to treat asthma, carbuncle, cholera, colic, constipation, convulsion, delirium, head-ache, paralysis, post-natal complaints, syphilis, tumours and small-pox. Plant is useful as antidote on snake-bite, also used as veterinary medicine in various diseases. Plant is eaten by cattle as fodder.

(Jain, 1991)

Critical Notes

Paste of the plant is useful to treat injuries in case of animals. Medicinal utility is already reported. The present report increases authenticity of its applicability. Use of seeds against devil's power and for conception aer not reported earlier.

**Dolichandrone falcata* Seem.

Family: Bignoniaceae, Local Name: Modshing, Medshing

Ethnobotanical Notes

Veterinary Medicine

1. Decoction of stem bark is administered to hens against dysentery.

Notes from Classics

Wood is used for making agricultural implements. Fruits are tied on waist of a bridegroom at marriage in Hindus.

(Watt, 1889–1893)

Stem bark is used as fish-poison. Stem bark yields coarse fibre.

(Anonymous, 1948–1976)

Bark is used as fish poison. Bark yields coarse fibre. Wood is used for house construction and agricultural implements.

(Ambasta, 1986)

Plant is said to be abortifacient. Leaves are used to treat body-ache. Leaves are used against snake-bite.

(Jain, 1986)

Critical Notes

Stem bark is helpful to treat dysentery in case of hens. It is not noted in the classical literature.

Dolichus lablab L.

Family: Papilionaceae, Local Name: Val

Ethnobotanical Notes

Miscellaneous

1. Leaves constitute a good fodder for goats, bullocks, buffaloes etc.
2. Pods and seeds are cooked and eaten as vegetable.

Notes from Classics

In Assam, roots are used to poison wild animals. Bean juice mixed with salt is applied on inflation of ear and throat due to cold. Stem is used as cattle fodder.

(Watt, 1889–1893)

Seeds are febrifuge, stomachic and antipasmodic. Bean is popular throughout India. Plant is used as goat fodder.

(Anonymous, 1948–1976)

Ambasta (1986) gave uses similar to those in Anonymous (*loc.cit.*).

Roots are used as a vermicide and also help in antifertility. Leaves are said to be useful on cough and to treat ring-worms. Pods and seeds are cooked as vegetable. Leaves are used as green dye for tattoo.

(Jain, 1991)

**Echinops echinatus* Roxb.

Family: Compositae, Local Name: Udkata

Ethnobotanical Notes

Medicine

1. Roots are same plant, and roots of *Dichrostachys cinerea* (L.) Wight and Arn., *Dolichos trilobus* L. are crushed together in water and mixture is kept over night. Next morning, infusion is given to control dysentery in children. Two to three spoons are advised per day till cure.

2. Roots are chewed on urinal complaints.
3. Roots are rubbed on stone and paste is applied on the place of scorpion sting.

Veterinary Medicine

1. Root extract is applied on body of animal as a remedy for ticks.

Notes from Classics

Root powder mixed with gum of *Acacia nilotica* is applied on hair as a remedy for lice. Entire plant is considered as bitter tonic and diuretic. It is useful in hoarse cough, hysteria, dyspersia, scrofula, and also used in opthalmia. Root powder is applied on wounds of cattle to get rid off maggots.

(Anonymous, 1948–1976)

Ambasta (1986) mentioned uses similar to those in Anonymous (*loc.cit.*).

Roots are useful in the treatment of skin diseases and bone-ache. Roots are used to facilitate child-birth. Roots are used as veterinary medicine for various diseases. Entire plant is useful to treat malaria and fever.

(Jain, 1991)

Critical Notes

Roots are variously useful as noted above by the present investigators. All these are not noted in past.

**Eclipta alba* (Linn.) Hassk.

Family: Compositae, Local Name: Kala maka

Ethnobotanical Notes

Medicine

1. Leaf juice is applied on foot to cure 'Chikhali', an injury caused due to constant contact of mud in rainy season.

Notes from Classics

Fresh root juice is purgative and employed to anaemic persons. Plant juice is applied on head of infants to blacken hair. Plant exract is useful internally or externally to blacken hair on head. It is the oldest Hindu medicine as tonic and deobstruent in hepatic and spleenic enlargements and various chronic skin diseases. Plant juice in combination with aromatics such as 'Ajowan' seeds, as a tonic and deobstruant and give two drops of it with honey to children suffering from catarrah. It forms and ingredient of a remedy used for tetanus is Konkan. Herb oil is mixed with sesamum oil, externally applied on elephantiasis. Root juice is applied on gall on neck of cattle.

(Watt, 1889–1893)

Leaf juice mixed with honey is given as a remedy for catarrah in infants. Plant juice is administered in combination with atomatics for cattarrahal jaundice. Herb is used as tonic and deobstruant in hepatic and spleen enlargements and for skin diseases. Fresh plant is rubbed on gums for tooth-ache and also applied on head with little oil for relief from headache. Plant juice mixed with sesamum oil is used on elephantiasis. Roots paste is applied externally as antiseptic, also used on ulcers and wounds of cattle. Leaves are used as vegetable in Java and made into chutneys in India.

(Anonymous, 1948–1976)

Root is said to be emetic and purgative. Plant is considered as tonic, deobstruant, used in enlargement of spleen and skin diseases. Root paste is applied on wounds of cattle.

(Ambasta, 1986)

Root juice is used as tonic. Leaf juice is used as tonic. Leaf juice is considered as antiseptic, use of leaf juice is considered as antiseptic, used on carbuncle. It is cooling, used on eye diseases like conjunctivitis, malaria and fever, also useful in gastric complaints, promotes hair growth, cures skin diseases, sores and swellings of ears. Flower buds are used against headache. Entire plant is useful against asthma, bronchitis, gland swellings, jaundice, liver complaint, leucoderma, toothache, ulcers, hepatic disease, itch and wounds. Roots are used as antidote on snake-bite. Leaf juice is dropped on scorpion bite.

(Jain, 1991)

Critical Notes

Use of leaf juice to cure 'chikhli' is new.

Ehretia laevis Roxb.

Family: Ehretiaceae, Local Name: Lokhandi

Ethnobotanical Notes

Miscellaneous

1. Ripe fruits are edible.

Notes from Classics

Leaves are used as cattle fodder. Wood is used for building purposes. Fruits are eaten in times of famine.

(Watt, 1889–1893)

Anonymous (1948–1976) and Ambasta (1986) recorded similar uses those recorded by Watt (*loc.cit.*).

Fruits, flowers and inner bark are edible. Leaf is used on muscle pain. Stem is used for thatching huts. Entire plant is used on syphilis.

(Jain, 1991)

Eleusine coracana (L.) Gaertn.

Family: Graminae, Local Name: Nagali

Ethnobotanical Notes

Medicine

1. Grain flour is baked into cakes or its paste is advised for consumption especially after delivery. It keeps good health of a mother.

Notes from Classics

Grains are highly nourishing, used as diet for prisoners and for poorer classes doing labourous work. Fermented liquor called 'bajah' or 'bajali' is prepared from grains which is similar to beverage. Straw is excellent fodder for cattle.

(Watt, 1889–1893)

It is principal food grain of agricultural classes, grains ground into flour is used to prepare cakes, puddings and porridge etc. A fermented drink or beer is prepared from grains. Ragi is recommended wholesome food for a diabetic patient. Ragi straw is a good fodder for cattle.

(Anonymous, 1948–1976)

Ambasta (1986) recorded uses similar those in Anonymous (*loc.cit.*).

The entire plant is used on measles, pneumonia and small-pox. Grain is useful in skin eruptions. Grain flour is roasted and cakes are edible. Grains are fermented and used as local drink.

(Jain, 1991)

Critical Notes

The present report reiterates the nourishing value of the grains especially when a woman needs good diet.

**Elytraria acaulis* (L.*f.*) Lind.

Family: Acanthaceae, Local Name: Vismuli

Ethnobotanical Notes

Medicine

1. Roots are chewed for urinal complaints.

Notes from Classics

Infusion of entire plant is used as a cough remedy for infants. Decoction of leaves is used for veneral diseases.

(Watt, 1889–1893)

Roots are applied on boils and burns, also used on colic, diarrhoea, hardness of breast tissues, rickets, throat complaints and tonsils. Leaves are administered on ear-ache, fever, nail diseases and applied on wounds. Whole plant is used on abscess of mammary glands. Leaves are useful on veterinary sores.

(Jain, 1991)

Critical Notes

Application of roots against urinal complaints is additional one.

***Emblilca officinalis* Gaertn.**

Family: Euphorbiaceae, Local Name: Awala

Ethnobotanical Notes

Medicine

1. Stem bark is crushed and kept is water. A glass of infusion is given as antidote against snake-bite, which results in vomiting and reduces poison.

2. Fruits are boiled in sugar water and cooled, given to persons suffering from jaundice.

Miscellaneous

1. Ripe fruits are edible.

Notes from Classics

Root bark is astringent. Fruits are acrid, cooling refrigerant, diuretic and laxative. Raw fruits are eaten as aperient. Dried fruits are useful in haemorrhage, diarrhoea and dysentery. In combination with iron, it is used as a remedy for anaemia, jaundice and dyspepsia. A liquor prepared from fruit is useful in jaundice, dyspepsia and cough. Acute bacillary dysentery may be arrested by drinking a sherbet of amla with lemon. Exudation from incisions on the fruit is used as external application for inflammation of eyes. Flowers are cooling refrigerant and aperient. Seeds are used in the treatment of asthma, bronchitis and biliousness. Leaves are used as fodder and green manure. Wood is used for agricultural implements, building purposes and furniture.

(Anonymous, 1948–1976)

Fruits are sour and astringent, cooling, diuretic, laxative, eaten raw or cooked, also pickled. Fruits are used in hair dyes, dried ones are detergent and used for shampooing hair. Wood is used for agricultural implements, poles and furniture.

(Ambasta, 1986)

Stem bark powder is applied on burns. Leaves are useful as a remedy on cold, dysentery, and applied on wounds. Leaves are used on scorpion bite. Fruits are anthelmentic, used to treat asthma, bronchitis, constipation, diabetes, diuretic, used on dysentery, also used in the treatment of eye complaint, headache, fever indigestion. They are laxative, used on liver complaints, and as purgative, used on stomach complaints, thirst, as tonic and also used on vaginal complaint. Seeds are used on madness, and also used on asthma. Leaves are used on veterinary stomach diseases. Fruits are pickled and eaten raw, and used in local drinks. Tree is abode of God Vishnu. Tree are believed born from tears of God. Tulsi is married with this free then the weddings follow Tree is offering to Lord Vishnu and Lord Shiva.

(Jain, 1991)

Critical Notes

Uses of stem bark against snake-bite and fruits to treat jaundice are new applications.

**Emilia sonchifolia* (L.) DC.

Family: Compositae, Local Name: Dhampan

Ethnobotanical Notes

Medicine

1. Decoction from entire plant is useful in asthma. 15-20 ml is administered twice a day for a week.

Notes from Classics

Roots are used to treat diarrhoea. Fresh juice of roots is used on sore eyes, ears and night blindness. It is also used on cuts and wounds. Decoction of plant is used as febrifuge in infantile tymphanites and in bowel complaints.

(Anonymous, 1948–1976)

Leaf juice is used on sore eye and ears. Decoction of herb is used as febrifuge and also in bowel complaints. Leaves are cooked and eaten as vegetable or as salad.

(Ambasta, 1986)

Root bark is used to treat diarrhoea. Leaves are used as application on boils, bruises, cataract, eye complaints, inflammation and night blindness. Root in combination with leaves is useful in the treatment against gangrene. Entire plant is useful in bowel complaints, and is said to be cooling. Leaves are cooked and eaten as vegetable, or also eaten as salad.

(Jain, 1991)

Critical Notes

Plant is beneficial in case of asthmatic patients. This is not documented yet.

**Ensete superbum* (Roxb.) Cheesm.

Family: Musaceae, Local Name: Rankela

Ethnobotanical Notes

Miscellaneous

1. The core of inflorescence axis is used as vegetable

Veterinary Medicine

1. Seeds are crushed in water and then given orally to chicken for dysentery.

Notes from Classics

Inflorescence and fruits are edible. Inflorescence are also pickled.

(Watt, 1889–1893)

Flower buds are used as vegetable. Young shoots and inflorescence are cooked and eaten as vegetable. Fruits are eaten raw.

(Jain, 1991)

Critical Notes

Medicinal utility of seeds to check dysentery in case of chicken is yet unreported.

**Eranthemum roseum* Linn.

Family: Acanthaceae, Local Name: Mortutiya

Ethnobotanical Notes

Veterinary Medicine

1. Roots are fed mixed with fodder in case of milking animals.

Notes from Classics

Tuberous roots are boiled in milk and administered to check leucorrhoea. Roots are also helpful to cows and buffaloes to promote growth of fetus.

(Anonymous, 1948–1976)

Ambasta (1986) mentioned uses as those in Anonymous (*loc.cit.*).

Roots are useful on small pox. Roots are used as galactogogue.

(Jain, 1991)

**Eriophorum comosum* Wall.

Family: Cyperaceae, Local Name: Molgavat

Ethnobotanical Notes

Miscellaneous

1. Entire plant is used for making brooms to sweep houses and foreground.
2. Plants are used to make rope which is used to tie bundles of fodder and fire-wood.

Notes from Classics

Plant yield fibre for ropes.

(Watt, 1889–1893)

Plant yields material for ropes but inferior to 'Bhabhar' grass (*Elaliopsis binata* C.E. Hubbard).

(Ambasta, 1986)

Whole plant is used for cordage.

(Jain, 1991)

Critical Notes

Use of the plant for making brooms is new.

**Erythrina stricta* Roxb.

Family: Papilionaceae, Local Name: Pangara

Ethnobotanical Notes

Medicine

1. Leaf juice of *E. stricta* and *Butea monosperma* L. is given to

treat cough. Two spoons alongwith honey is advised orally for 3-4 days.

Miscellaneous

1. Wooden riffs are used to make shutters of cattle sheds.

Notes from Classics

Wood is used for making planks.

(Watt, 1889–1893)

Bark powder is used in the treatment of biliousness, fever, rheumatism, asthma, leprosy, itch and epilepsy. Flowers are given as antidote to snake-poison.

(Anonymous, 1948–1976)

Bark is pulverized and is used on biliousness, rheumatism, asthma, leprosy and epilepsy. Flowers are useful as antidote against snakebite. Wood is useful in making scabbards, ladders, ghee storing jars, drums, water troughs, packing cases, and match boxes etc.

(Ambasta, 1986)

Stem bark is useful on fever, menstrual complaints and stomach diseases. Leaves are used as tonic and also used as vermifuge. Stem bark is used as antidote on snake bite.

(Jain, 1991)

Critical Notes

Leaf juice is unreported remedy against cough. Wood being light is used for making door shutters of cattle sheds. This is also not documented.

**Erythrina suberosa* Roxb.

Family: Papilionaceae, Local Name: Darara, Pangara

Ethnobotanical Notes

Miscellaneous

1. Stem bark is crushed and thrown in water bodies for fish killing.

Notes from Classics

Wood is used for scabbards, sieve frames, and occasionally for planking.

(Watt, 1889–1893)

Bark is used for bottle stoppers, crown cork liners and insulation boards. Timber is said to be lightest wood of India. Used for making ghee jars, scabbards, ladders, tea boxes, plywood and match boxes.

(Anonymous, 1948–1976)

Ambasta (1986) mentioned uses similar to those in Anonymous (*loc.cit.*).

Stem bark is used to treat dysentery. Leaves are useful to destroy maggots in ulcers.

(Jain, 1991)

Critical Notes

Utility of stem bark to stupefy fishes is unreported.

**Erythrina variegata* Linn.

Family: Papilionaceae, Local Name: Ghadhapalas, Chopade pangara

Ethnobotanical Notes

Veterinary Medicine

1. Stem bark is burnt and the ash obtained is mixed with coconut oil. This paste is applied on injuries caused by yoke on bullock's shoulder.

Notes from Classics

Stem bark is antibilous and a febrifuge is useful as collyrium in opthalmia. Leaves are applied externally to get relief from joint pains. Fresh juice of leaves is injected in ears against ear-ache and also as an anodyne in tooth-ache. Tender leaves are eaten in curry. Leaves are used as cattle fodder.

(Watt, 1889–1893)

Bark is astringent and used as febrifuge, antibilious and anthelmintic also useful on opthalmia. Leaves are laxative, diuretic, anthelmintic, glactogogue and emmenogogue. Leaf juice gives relief on ear-ache and used as anodyne on tooth-ache, also useful killing worms in sores. Leaves are externally applied on joint pains. Leaves and tender shoots are used as pot-herb. Green leaves are used as cattle fodder, also used as green manure.

(Anonymous, 1948–1976)

Leaves are considered to be laxative, diuretic, anthelmintic, galactogogue, emmenogogue and also applied externally on veneral buboes. Bark yields fibre used for making cordage. Wood is used for rafts, floats, canoes and catarmans, also suitable for paper pulp.

(Ambasta, 1986)

Stem bark is antihelmentic, used on cold, fever, eye complaints and rheumatism. Leaves are useful on ulcers. Entire plant is useful on mennorhoea, pimples and paralysis. Fruits are cooked and eaten as vegetable. Plant is used as antidote to snake-bite.

(Jain, 1991)

Critical Notes

Application of stem bark for ethnoveterinary purpose is still not reported.

Eulophia herbacea Lindl.

Family: Orchidaceae, Local Name: Umarkand

Ethnobotanical Notes

Medicine

1. Decoction of tuber is used on spermatorrhoea urinary complaints and menses. Doses vary from two to four spoons and are generally advised at night after meals.

Notes from Classics

Tubers are source of salep.

(Ambasta, 1986)

Pseudobulbs are used as tonic.

(Jain, 1991)

Critical Notes

Tubers are beneficial to treat the aforesaid diseases. These are still unrevealed from other parts of our country.

**Euphorbia fusiformis* Buch.-Ham. ex D.Don

Family: Euphorbiaceae, Local Name: Bhuisherda

Ethnobotanical Notes

Medicine

1. Peels removed from tuberous stem axis are heated and applied on swellings of stomach at night.

Veterinary Medicine

1. Latex is applied on cut made at dorsal base of tail to cure infection of shoulder of bullocks.

Notes from Classics

Roots are administered on fever, gout and rheumatism. Roots are used as veterinary medicine on dysentery.

(Jain, 1991)

Critical Notes

All applications noted by the present investigators are new and noteworthy.

***Euphorbia geniculata* Orteg.

Family: Euphorbiaceae, Local Name: Dudhi

Ethnobotanical Notes

Miscellaneous

1. Leaves are fed as fodder for pet animals.

Critical Notes

The usefulness of plants as fodder is not noted in past.

**Euphorbia neriifolia* Linn.

Family: Euphorbiaceae, Local Name: Sabar

Ethnobotanical Notes

Medicine

1. Bark of young stem is scrapped and roasted, juice obtained is dropped in ears against ear-ache.

Miscellaneous

1. It is planted as live hedge for crop fields.

Notes from Classics

Roots mixed with black pepper are employed in cases of snake bite. Milky juice is considered as purgative and rubefacient. Milky juice is applied on ear-ache and if mixed with soot is employed on glandular swellings and to prevent suppuration. Tender branches are slightly roasted and juice is squeezed out and given with molasses for producing vomiting and purging in bronchitis of children. It is also applied on unhealthy ulcers and scabies. Pulp of stem is mixed with ginger and given to persons injured by mad dogs before the accession of hydrophobia. Shrub is sacred to 'Mansa', the Godesses of serpents. Plant is planted on 5th day of full moon in month of Shrawan in courtyard of Hindus and tree is worshiped.

(Watt, 1889–1893)

Latex is acrid rubefacient, purgative and expectorant. Plant juice is employed in ear-ache, mixed with soot is applied on opthalmia. Plant juice is also used as an excellent remedy for asthma.

(Anonymous, 1948–1976)

Latex is acrid, rubefacient, purgative and expectorant, used to remove warts, and cutaneous eruptions.

(Ambasta, 1986)

Critical Notes

Its use as live hedge is obvious, although not reported so in the past.

**Euphorbia parviflora* L.

Family: Euphorbiaceae, Local Name: Dhudhi

Ethnobotanical Notes

Miscellaneous

1. Plant is used as fodder for cattle.

Notes from Classics

Infusion of leaves is a remedy on dysentery, diarrhoea, menorrhagia and leucorrhoea. It is given with milk on colic for children.

(Watt, 1889–1893)

An infusion of dried leaves is considered astringent and feebly narcotic. It is used in dysentery, diarrhoea, menorrhagia and leucorrhoea. Infusion with milk is given to children in colic.

(Anonymous, 1948–1976)

Infusion is used as astringent in diarrhoea, dysentery and leucorrhoea. (Ambasta, 1986)

Leaves are used on diarrhoea, dysentery and leucorrhoea.

(Jain, 1991)

Critical Notes

Fodder value of this species is noted for the first time. It is a crop weed.

Ficus amplissima J.E.Sm.

Family: Moraceae, Local Name: Pipri

Ethnobotanical Notes

Miscellaneous

1. Leaves are used as cattle fodder.
2. Wood is useful as fuel.

Notes from Classics

Bark yields good fibre. Wood is used as fuel.

(Watt, 1889–1893)

Bark is used in colic. Leaves are used as cattle fodder. Wood is used for furniture and also as fuel.

(Anonymous, 1948–1976)

Ambasta (1986) recorded similar uses as recorded by Anonymous (*loc.cit.*).

Ficus arnottiana (Miq.) Miq.

Family: Moraceae, Local Name: Khadak-payer

Ethnobotanical Notes

Medicine

1. Latex is applied on pimples and pustules of body.
2. Latex is applied on boils caused due to plummet.
3. Latex is dropped on rice grains, after 2 to 3 hours are eaten by mothers to increase lactation. This is also similarly beneficial for cows.

Miscellaneous

1. Fruits are eaten raw.
2. Leaves are used as fodder for goats and cattle.

Notes from Classics

Leaves and bark are useful on cutaneous affections. Leaves are lopped for fodder.

(Ambasta, 1986)

Leaves are used in tea.

(Jain, 1991)

Critical Notes

The various applications of latex noted by the present investigators are all additional. Edibility of fruits is also a new information.

*Ficus bengalensis (Miq.) Miq.

Family: Moraceae, Local Name: Vad

Ethnobotanical Notes

Medicine

1. About 10 ml of latex is diluted in water and drunk to check yellow urination.
2. Decoction of aerial roots about two spoon is administered as hair tonic.

Miscellaneous

1. Aerial roots are splitted and used for tying bundles of crop fodder and fire-wood.

Notes from Classics

Root fibres are said to be useful on gonorrhoea. Infusion of bark is given as a powerful tonic for diabetic patients. Infusion of branches is useful in hoemoptysis. Tender ends of aerial roots are given for obstinate vomiting. Leaves are applied as poultice to abscesses, and decoction of yellow leaves and roasted rice is used as diaphoretic. Young buds are given in diarrhoea. Milky juice is externally applied on bruises and as an anodyne, also used on rheumatism, lumbago, foot cracks or inflammations, also applied on gums as a remedy for tooth-ache. Seeds are deemed cooling and tonic. Small quantity of milky juice is given to control dysentery. Leaves and twigs are greedly eaten by elephants. Leaves are used as platters. Small red figs are eaten by poorer classes. Wood is used for boxes, tent-poles, cart yokes etc. Bark of aerial roots yield fibre used for rope making. Milky juice is used to make bird-lime by natives.

(Watt, 1889–1893)

Anonymous (1948–76) and Ambasta (1986) mentioned uses similar to those in Watt (*loc.cit.*).

Aerial roots are used in the treatment of leucorrhoea, menorrhagia and also used as tooth brushes for cleaning teeth. Bark is useful in diabetes and urinary complaints. Leaves are used as remedy on cough, blisters and boils. All parts of plant in combination are used in the treatment of atrophy, dysentery, fever, tooth-ache, scabies and sores in mouth. Latex is applied externally on blisters and boils, eye complaint, skin diseases, spermatorrhoea, also used as tonic. Ripe fruits are edible. Tree is symbolically sacred, abode of Shankar and Savitri roots abode of God Bramha, trunk for Vishnu, and branches from Shankar. Cutting tree is more sinful than killing saint. Tree is thought created by Shiva, clan name 'Barle' in central India based on this. Raja Satyawana died under this tree. Savitri, his wife, brought him back to life. Huge tree on Narmada bank is called 'Kabirbghat'.

(Jain, 1991)

Critical Notes

Applying latex against yellow urination is additional. Roots are beneficial as hair tonic. These find no place in classical literature.

**Ficus glomerata* Roxb.

Family: Moraceae, Local Name: Umbar

Ethnobotanical Notes

Medicine

1. Root latex is collected without cutting root from mother plant by digging a pit. One spoon latex is given thrice a day for one week to person as a remedy for chicken-pox and small-pox.
2. Latex is mixed with sugar and given thrice a day for three days on stomach-ache and bleeding excreta.
3. Latex is applied on naval part as a remedy for stomach-ache.
4. Latex is applied on gums to strengthen them.
5. Latex is applied on swollen body parts.
6. Latex is applied in case of nipple ache.
7. Stem is used as tooth brush.

Miscellaneous

1. Leaves are fed to goats and cattle especially during summer months.
2. Ripe receptacles are cherished by all.
3. Wood is used as fuel.

Notes from Classics

Bark is given as astringent wash for wounds, also used to release poison of tiger or cat bite. Roots are useful in dysentery, fluid obtained from roots is administered as a powerful tonic. Leaf powder mixed with honey is given on bilious affection. Leaves are soaked in milk, mixed with honey is given to prevent from small pox. Twigs are given in menorrhagia and haemoptysis. Milk juice from stem is administered on piles and diarrhoea. It is useful with combination of sesamum oil is on cancer. Fresh juice is fruit is given on diabetes and urinary complaints. Sap is said to be used on mumps and other inflammatory and glandular enlargements. Fruits are eaten raw or cooked in curries.

(Watt, 1889–1893)

Fruit powder forms a valuable breakfast food, similar to imported grape nuts. Leaves are used as fodder for cattle and elephants. Latex is used in manufacture of ground sheets and water proof powdered paper. Wood is used for well curbs, house doors, rice mortars, shutter, carts, furniture, toys and agricultural implements etc.

(Anonymous, 1946–1976)

Roots are useful in diarrhoea and diabetes. Leaves are used on bilious affections. Fruits are stomachic and carminative, used in haemoptysis. Latex is used in the treatement of piles and diarrhoea and also used as bird-lime. Bark paste is used as vulnerary. Fruits are roasted and pounded, used as a break -fast. Wood is used for making well curbs, furniture, fuse box, match boxes etc.

(Ambasta, 1986)

Roots are used on antifertility, veneral complaints and dysentery. Stem is used as tooth brush. Bark is used in the treatment of cancer. Leaves are used to treat pneumonia. Fruits are carminative, used as galactogogue, on diabetes, stomach-ache and urinary complaints. Latex is used on blisters, boils, diarrhoea, dysentery, piles, pulmonary discharge in children, skin cracks on heels and lips, also used for joint dislocations. Plant is useful on hydrocele, scabies, menustrual complaints, muscle pain, spermatorrhoea and small pox. Tree is symbolically sacred; to make effigy corpse lost or not available. Bark is used as veterinary medicine for plague. Fruits are edible.

(Jain, 1991)

Critical Notes

Latex obtained especially from roots is beneficial in small-pox and chicken-pox. Latex from the plant is used in cases of bleeding in excreta, stomach-ache, swollen body parts and nipple-ache. All these applications are new.

**Ficus religiosa* Linn.

Family: Moraceae, Local Name: Pipal, Pimpal

Ethnobotanical Notes

Medicine

1. **Stem** bark paste is applied on skin infections, and also cures swellings.

Miscellaneous

1. Ripe fruits are eaten by tribals.
2. Leaves are used as fodder for cattle.

Notes from Classics

Bark is considered astringent, used on gonorrhoea. Infusion of bark is given internally on scabies. Paste of bark is employed on inflammatory swelling. Leaves and young shoots are administered on skin diseases. Powder of fruits is given to treat asthma. Powder taken with water is administered to woman to make her fruitful. Leaves and branches are used as fodder for goats. Small figs are cooked and eaten as vegetable by Gonda tribes. Fruits are helpful to improve digestion. Bark and leaves are rarely used for tanning bark, also yields fibre. Trees are planted as avenue trees and are sacred to Hindus. The roots are Brahma, the bark Vishnu and branches the Mahadeos.

(Watt, 1889–1893)

Leaves are purgative and used on skin diseases. Leaves are lopped for cattle and elephants as fodder. Fruits and tender buds are eaten by during scarcity. Latex is used as sealing wax to fill cavities in hallow ornaments.

(Anonymous, 1948–1976)

Infusion of bark is used on ulcers and skin complaints. Bark yields fibre which is used for paper making. Leaves are lopped for fodder. Fruits and tender buds are eaten in times of scarcity. Latex is used to fill hollow cavities of ornaments. It is also a source of bird-lime. Wood is used for packing cases, occasionally for felloes of wheels, spoons and bowels etc.

(Ambasta, 1986)

Bark is used as application for blisters, boils, ulcers, gonorrhoea, haemoptysis, scabies and whooping cough. Leaves are considered as abortifacient, used on boils, ulcers and skin diseases. Latex is used as tonic. Fruits are laxative and edible. Entire plant is used on carbuncle, dysentery, fever, rheumatism, small-pox and spermatorrhoea. Stems are burnt in 'Yagna' a taboo.

(Jain, 1991)

Critical Notes

Bark is noted for the first time to treat skin infections and swellings.

**Ficus virens* (Miq.) Miq.

Family: Moraceae, Local Name: Ambe-payer, Amchor

Ethnobotanical Notes

Medicine

1. Latex is applied for skin infections like eczema. Latex is used to cure boils caused due to uprooting hair.

Miscellaneous

1. Leaves are fed to goats.
2. Ripe receptacles are eaten by tribals.

Notes from Classics

Bark of same plant, with other four species of Ficus and bark of *Melia azadirachta* called 'Panchval kada' or the decoction is used for gargaling in case of ulcerations. Young shoots are eaten in curries by natives. Leaves are used as fodder for cattle and elephants.

(Watt, 1889–1893)

Decoction of bark is used as gargle and vulnerary. Fruits are used as fodder for cattle. Wood is suitable for paper pulp.

(Ambasta, 1986)

Fruits are buds are edible. Young leaves are cooked as vegetable. Shoots are pickled.

(Jain, 1991)

Critical Notes

Applications of latex to treat eczema and boils caused due to uprooting of hair are not revealed elsewhere.

**Flacourtia indica* (Burm.*f.*) Merr.

Family: Flacourtiaceae, Local Name: Galguggar

Ethnobotanical Notes

Miscellaneous

1. Leaves are fed to goats as fodder.
2. Ripe fruits are eaten by tribals.

Notes from Classics

In Malaya, shoots and leaves are pounded with coconut milk and lotion is used for washing hair. Fruits are used in the treatment of small-pox. Split stems are used to tie and also employed for making baskets.

(Anonymous, 1948–1976)

Roots are used on skin diseases. Bark is prescribed against dysentery, eczema, rheumatism, gout and also facilitates child-birth. Fruits are eaten to improve digestion and also a remedy for jaundice. Plant is used in the treatment against burning sensation in chest, carbuncle, cholera, fever, liver diseases and wounds. Ripe fruits are edible. Bark is used in the treatment against mad-dog bite. Roots are used as antidote against snake bite. Plant is used as veterinary medicine on foot and mouth diseases of cattle.

(Jain, 1991)

Critical Notes

Fodder value of the plant is not noted earlier.

**Garuga pinnata* Roxb.

Family: Burseraceae, Local Name: Kankad

Ethnobotanical Notes

Miscellaneous

1. Fruits are crushed and spread in water bodies to kill fishes.
2. 'Haal' a horizontal container of water to drink for chicks is made from the wood.

Notes from Classics

Juice extracted for the stem is dropped in eyes as a remedy for opacities and conjunctiva. In Konkan, juice of leaves, alongwith the leaves of Vasica and *Vitex trifolia* are mixed with honey and given as a treatment of asthma. Fleshy drupes are eaten raw, pickled or cooked by natives. Shoots and leaves are collected as fodder for elephants. Bark is used for taning in many parts of country.

(Watt, 1889–1893)

Stem juice is used as a remedy for opacity and conjunctiva. Leaf juice along with honey is given for asthma. Leaves and shoots are used as fodder. Bark and leaf galls are used for tanning. Wood is used as fuel. Drupes are eaten raw. Timber is used for planking, canoes, boxes, drums, cabinet work, house holdings, wood also suitable for commercial tea chest, plywood, match splints and pencils etc.

(Anonymous, 1948–1976)

Ambasta (1986) cited uses similar to those in Anonymous (*loc.cit.*).

Roots are used for curing skin diseases and veneral complaints. Bark is used to treat diabetes, eye complaints and wounds. Leaves are considered as astringent, are used on asthma. Fruits are used against dysentery. Fruits are edible. Bark is used as fish poison for trapping fish, also used as black dye for dyeing purposes.

(Jain, 1991)

Critical Notes

Application of fruits to stupefy fishes is undocumented yet.

**Geodorum densiflorum* (Lam.) Schlect.

Family: Orchidaceae, Local Name: Harya-kand

Ethnobotanical Notes

Medicine

1. Rhizome paste is applied on goitre.
2. Rhizomes are consumed along with honey for two to three months to treat impotency.
3. Rhizomes are rubbed on skin to cure swellings.

Notes from Classics

Roots stocks are crushed and rubbed on body of cattle to get rid off flies and gadflies.

(Anonymous, 1948–1976)

Root stock contains a gummy substance, and are employed in the preparation of glue used for joining the parts of musical instruments.

(Ambasta, 1986)

Critical Notes

Medicinal utility of rhizome on goitre, impotency and swellings is hitherto not reported.

Gliricidia sepium (Jacq.) Walp.

Family: Papilionaceae, Local Name: Gulchhadi

Ethnobotanical Notes

Miscellaneous

1. Flowers are used to kill rats in houses.
2. Branches are used for making partition walls and thatching of roofs of huts.

Notes from Classics

Leaves are used as cattle fodder. Flowers are used as vegetable in Philipines. Powdered seeds, leaves and bark are mixed with rice and used as bait for destruction of pests.

(Anonymous, 1948–1976)

Leaves are lopped for fodder, fed to cattle. Flowers are used as vegetable. Wood is used for house posts, fences, stakes and railway cross-ties.

(Ambasta, 1986)

Critical Notes

It is exotic, native of Central America. It is planted especially on waste lands as a part of government afforestation programme in this region. Although it is introduced recently tribal people of Nasik district has tried to use them for aforesaid purpose.

**Gloriosa superba* L.

Family: Liliaceae, Local Name: Kalva, Kal-lavi

Ethnobotanical Notes

Medicine

1. Tubers are mixed in fodder to increase lactation.
2. Tubers are rubbed on stone and slurry is given for easy delivery.

Miscellaneous

1. Flowers are collected and sold in market at Trimbakeshwar as an offering to Lord Shiva.

Notes from Classics

Root paste is applied to naval supra pubic region of vagina to promote labour. Starch obtained from roots is given as a remedy on gonorrhoea. Roots are considered as tonic and stomachic. When necessary one to two roots are effective as an antidote on cobra-bite, scorpion sting and centepede sting.

(Watt, 1889–1893)

Tubers are regarded as tonic and stomachic. Tubers are administeted to cattle to expel worms. Leaf juice is applied on hair to get rid off lice.

(Anonymous, 1948–1976)

Tubers are regarded as tonic, stomachic and anthelmintic. Tubers are also used to promote labour pain and as an abortifacient, externally used for neuralgic pains and to cure skin complaints. Alkaloid colchicine is extracted from tubers used to treat gout and rheumatism.

(Ambasta, 1986).

Rhizomes are medicinally used as abortifacient, used in intermittent fever. They are useful in cholera, infertility, and facilitates child-birth, also used to treat gonorrhoea, gout, leprosy, piles, septics, ulcers, skin diseases, small-pox, stomachache and various other diseases. It is also used as tonic. Symbolically stem pieces are tied on arm of women to facilitate delivery. Leaves are used to treat asthma. Rhizomes are used as antidote on snake-bite, scorpion bite, also used as anthelminthic. Plant is useful on sores, spleen complaint, syphilis and rheumatism.

(Jain, 1991)

Critical Notes

Use of tubers to increase lactation in case of pet animals is still unreported. Flowers are offered to Lord Shiva at Trimbakeshwar. Either of the uses have greatly reduced the natural populations in the district. A stringent action is needed to conserve this taxon.

**Glossocardia bosvallea* (L.*f.*) DC.

Family: Compositae, Local Name: Mirgi kand

Ethnobotanical Notes

Medicine

1. Pieces of underground root-stocks are boiled in water alongwith sugar and tea powder and the decoction obtained is given to drunkers to get rid off the effect of alcohol drinking.

Notes from Classics

Leaves are used as vegetable. Plant is used on female complaint.

(Watt, 1889–1893)

Plant is used as vegetable in times of scarcity. It is also used as emmenogogue.

(Anonymous, 1948–1976)

Ambasta (1986) recorded uses similar as given by Anonymous (*loc.cit.*).

Critical Notes

To encounter the effect of alcohol drinking, root-stock is useful. This is not noted in classical literature. It is exotic and denizen of West Indies.

Gmelina arborea Roxb.

Family: Verbenaceae, Local Name: Shivan

Ethnobotanical Notes

Miscellaneous

1. Leaves are used as fodder for goats, cows and buffaloes.
2. Fruits are cherished by goats.
3. Wood is used for house construction.

Notes from Classics

Roots are recommended on fever, indigestion, anasarca and other complaints. Young leaf juice is used as demulcent in gonorrhoea and cough. Fruits are eaten by 'Gond' tribes. Leaves are used as fodder for cattle. Timber is used for making furniture, frames boxes etc.

(Watt, 1889–1893)

Decoction or infusion of roots is prescribed on indigestion, fever and anasarca. Roots are ingradients of Ayurvedic preparation 'Dasmula'. Leaf paste is applied on forehead as a remedy for fever and headache. Tender leaves are demulcent. Leaf juice is used as a wash for foul ulcers. Drupes, leaves, flowers and roots are used as medicine. Decoction of drupes is used on fevers and bilious affection. Timber is used for making furniture, carriages, printing blocks, carving, musical instruments, jute bobbins etc., also employed bridges, ship building, plywood and paper making.

(Anonymous, 1948–1976)

Leaves are demulcent. Decoction of fruits is used on fever and bilious affection. Wood is used for furniture, venation blinds, tight cooperage, carving, musical instruments, shafts, axles and various articles, also used for ship building, plywood, paper and match sticks manufacture.

(Ambasta, 1986)

Roots are helpful for blood purification, catarrah of bladder, malaria, also used as tonic. Leaves are used as application on gonorrhoea, headache and to cure wounds. Stem bark is used to treat stomach diseases, bone fractures etc. Stem bark is used as antidote against snake bite. Entire plant is used in anasarca, asthma, bronchitis, cholera, colic, diarrhoea, dropsy, dyspepsia, epilepsy, fever, rheumatism, small pox, sores, spleen complaints, syphilis, throat swelling and utricaria. Wood is used in preparation of combs, drums and for house construction.

(Jain, 1991)

**Grewia abutilifolia* Vent. ex A. Juss.

Family: Tiliaceae, Local Name: Chikan Kharbat, Chikan Kharata

Ethnobotanical Notes

Medicine

1. Root juice is applied on swellings.
2. Root paste is useful in diabetes.

Notes from Classics

Roots are useful on abscesses. Fruits are edible.

(Jain, 1991)

Critical Notes

Roots are useful in treating swellings and diabetes. These applications are not reported in past.

**Grewia tenax* (Forssk.) Fiori.

Family: Tiliaceae, Local Name: Kharbat, Ghangudi

Ethnobotanical Notes

Medicine

1. Root paste is applied on dislocation of bones in case of human beings.

Veterinary Medicine

1. Root paste is also useful for dislocation in case of pet animals.

Miscellaneous

1. Ripe Fruits are cherished by natives.

Notes from Classics

Plant is used as fodder for camels and goats. Fruits are edible. Wood is used as walking stick.

(Ambasta, 1986)

Stem extract is used to treat cough. Fruits are edible.

(Jain, 1991)

Critical Notes

Application of root paste for dislocations for both cases is new.

Grewia tiliaefolia Vahl

Family: Tiliaceae, Local Name: Sonjaman, Dhaman, Rodgi

Ethnobotanical Notes

Miscellaneous

1. Ripe fruits are edible.
2. Leaves are used as goat fodder.
3. Bark yields fibres to tie bundles of fodder and fuel-wood.

Notes from Classics

Bark after removal of suber is rubbed down with water and thick muscilage is mixed with flour of *Panicum miliacium* and given as a remedy on dysentery. Bark is employed to remove irritation in cow-itch. Small drupes are eaten by natives. Leaves are used as cattle fodder. Bark yields fibre used for making cordage and paper. Wood is used for shafts, shoulder poles, handles, fishing rods etc.

(Watt, 1889–1893)

Stem bark is acrid and employed in dysentery. Bark is applied externally to remove irritation from cow-itch. Wood powder used as an antidote to opium poisoning. Fruits are edible. Bark yields fibre used for cordage. Leaves are used as cattle fodder. Leaves are employed as a soap substitute for washing hair. Timber is used for agricultural implements, carts, carriages, frames, tools, shafts etc. It is also suitable for ornamental furniture, used in textile mill for shutter, rabbet tubes, bobbins, etc. It is also used for cooperage, golf shafts, billiard cues shafts, stumps, and bails etc.

(Anonymous, 1948–1976)

Bark is used in dysentery. Bark yields fibres useful for cordage. Fruits are edible. Wood is used for poles, shafts, frames, panels carriages, furniture, billiard cues, cricket stumps etc.

(Ambasta, 1986)

Roots are used in the treatment of syphilis. Bark is used for curdling milk. Leaves are eaten as vegetable. Fruits are eaten raw.

(Jain, 1991)

Guizotia abyssinica (L.*f.*) Cass.

Family: Compositae, Local Name: Khursani

Ethnobotanical Notes

Miscellaneous

1. Seed powder is used for making curries and vegetables.
2. Seed cake fed to cattle.
3. Flowers are offered to Lord Mahadeva at Trimbakeshwar.

Notes from Classics

Black achenes yield edible oil used for culinary purposes. After extraction of oil from seeds oil cakes are fed to milking cows. Oil is also used for lighting and painting purposes.

(Watt, 1889–1893)

Seeds are used for making chutneys and condiments. Oil is extracted from niger seed is used for edible purposes. Seed cake is utilized for feeding cattle, and also used as manure.

(Anonymous, 1948–1976)

Seed oil is used for edible purposes. Seed oil is used in manufacture of paints, soups, and is a adulterant in Rape seed and Sesame oil. Oil is also useful for illumination of lamps. Seed cake is fed to cattle, and also used as manure.

(Ambasta, 1986)

Seeds are used in the treatment of rheumatism. Seed oil is used for edible purposes.

(Jain, 1991)

Critical Notes

The plant species is cultivated in the district as a source of oil seeds. It is used in vegetable or 'chutneys'. Offering of flowers to Lord Mahadeva is noted in this region. It is exotic, native of Abyssimia.

**Haldina cordifolia* (Roxb.) Ridsd.

Family: Rubiaceae, Local Name: Hedh, Haldu

Ethnobotanical Notes

Medicine

1. Stem bark extract is heated and applied at night on aching back and waist till cure.
2. Stem bark of this plant, and of *Madhuca longifolia* (Koenig.) Macbr. *Bombax ceiba* Linn., *Schleichera oleosa* (Lour) Oken. and *Chrysopogon fulvus* (Spr.) Chiov, are crushed together. They are dried, burnt, and ash called 'Misri' is applied on burns.

Notes from Classics

Stem bark is antiseptic, used in the treatment of eye complaints. Paste of bark is applied on scars on skin and yellowing of body, to cure urinal complaints and worms in sores. Leaves are useful to treat cough, cold, headache and to cure wounds. Stem is used as veterinary medicine. Plant is used against night blindness, carbuncle and cholera.

(Jain, 1991)

Critical Notes

Application of stem bark in aching back is unnoticed. Alongwith other barks it is useful for treating burning. It is also not reported.

**Hamiltonia suaveolens* Roxb.

Family: Rubiaceae, Local Name: Narkhi

Ethnobotanical Notes

Medicine

1. Stem bark is boiled in water and vapours are allowed over the body of a person suffering from fever and also in case of anaemic persons.
2. About 20 ml of stem bark decoction is given to patient suffering from flatulence. It is administered once a day till cure.

Veterinary Medicine

1. Stem bark decoction is administered to cattle to increase appetite.

Notes from Classics

Leaves are used as fodder for buffaloes. Wood is used for making charcoal and gun powder.

(Watt, 1889–1893)

Roots are useful in diarrhoea. Bark is ground and rubbed on body is puerperal fever. Infusion of root is given in courbature. Plant is used alongwith mustard oil as application for wounds. Wood is used for making charcoal and gun powder. Leaves are fed to buffaloes.

(Anonymous, 1948–1976)

Roots are prescribed on diarrhoea and courbature. Wood is used for charcoal and gun powder.

(Ambasta, 1986)

Roots are used for treating ulcers and wounds.

(Jain, 1991)

Critical Notes

Stem bark is employed for anaemic condition, flatulence and to increase appetite. All these application are hitherto unreported in past.

***Haplanthodes verticillatus* (Roxb.) R.B.Majumdar

Family: Acanthaceae, Local Name: Ekharu

Ethnobotanical Notes

Miscellaneous

1. Leaves are put in drinking water or thrown in ponds or lakes to settle contaminations.

Critical Notes

The plant is being reported useful for the first time.

Hardwickia binata Roxb.

Family: Caesalpiniaceae, Local Name: Anjan

Ethnobotanical Notes

Miscellaneous

1. Leaves are fed especially to milking cows and buffaloes.
2. Wood is used for raising abodes and making ploughs and cultivators.
3. Wood is used as fuel.

Notes from Classics

Bark yield fibre for cordage and paper making. Leaves are given as fodder to cattle. Wood is used for bridges, house constructions, ornamental work, carts, tent pegs and agricultural implements.

(Watt, 1889–1893)

Bark yields fibre used for ropes and other agricultural purposes. Leaves are used as cattle fodder and green manure. Wood is used for cart wheels, ploughs, clod crushers, post and beams, mine props, bridges, tool handles, carvings etc.

(Anonymous, 1948–1976)

Ambasta (1986) recorded uses similar to those given by Watt (*loc.cit.*).

Critical Notes

The species can be recommended especially for afforestation programmes in this region because of great demand for fodder and as it needs little care for growing.

Helianthus annuus Linn.

Family: Compositae, Local Name: Suryaphul

Ethnobotanical Notes

Medicine

1. Roots of this plant, *Azadirachta indica* A. Juss., *Pogostemon parviflora* Bth. and *Achyranthes aspera* L. are mixed and kept in copper vessel containing water. It is boiled until the water reduces half. This vessel is kept below the bed of a person suffering from flatulence.

Notes from Classics

Oil is expressed from seeds, and used for culinary and table purposes. Oil is chiefly used in candle and soap making. Oil cake is valuable food for cattle and poultry. Roasted seeds are eaten as nuts. Seeds are fed to sheep, pigs, poultry, pegions, rabbits etc. Leaves are said to be valuable fodder for cattle and horses. Fibres are obtained from stalks. Stalks are used as fuel and also used as a manure.

(Watt, 1889–1893)

Leaves are employed in the treatment of malarial fevers in Causacus. A tincture of flowers and leaves is recommended in combination with balsam for bronchitis. Seeds are diuretic and expectorant, used in bronchital, laryangeal and pulmonary affections, cough and cold. Seed oil is used for cooking and salad oil, for manufacture of margarine, bakery goods, paints and varnishes. Seed cake is fed to cows and poultry. Hull is employed as bedding material for cattle and poultry. Stem stalks are used as fuel and green manure. Stem ash is used as fertilizer.

(Anonymous, 1948–1976)

Seeds are considered as diuretic and expectorant. Seeds are consumed raw, roasted or salted. Seeds yield fatty oil used for cooking purposes, also used as semi-drying oil in paints and varnishes. It is a proteinous feed for live-stock. Flowers yield yellow dye for dyeing purposes.

(Ambasta, 1986)

Roots are used as a remedy on tooth-ache. Plant is useful on carbuncle, colic, diarrhoea, dysentery, dysuria, eye complaint, fever, menorrhagia, nose bleeding, post natal, sores, spleen complaints, bone fracture, whitlow and wounds. Plant is employed for veterinary hoof and mouth diseases. Plant is used as antidote on snake-bite and scorpion sting.

(Jain, 1991)

Critical Notes

Roots are used to treat flatulence. It is a cultivated for its valuable seeds. It is exotic, native of North America.

Helicteres isora L.

Family: Sterculiaceae, Local Name: Murudsheng

Ethnobotanical Notes

Medicine

1. Dried follicles are pounded and powdered, added in water, kept over night and this infusion is administered on stomach-ache, gas complaints and also to control dysentery in children. Two to three spoons of it are advised for adults once a day till cure. In case of infants, less than one spoon is administered once or twice a day till cure.

Notes from Classics

Capsule is chiefly employed in intestinal complaints, for colics, flatulence, and diarrhoea, also used on diabetes. Capsule powder along with castor oil is used as an application for discharge from ear. It is also employed for worms and colic in children. A fruit piece tied on body of newly born infant is believed to protect it from all kinds of bowel complaints. Leaves are used as fodder for buffaloes. Woody branches are used for thatching huts and also used as fuel.

Fruits of this plant and *Randia dumutorum* are tied on wrist of bride groom in marriage ceremonies of Vaisya caste.

(Watt, 1889–1893)

Root juice is used on empyema (collection of pus in body cavity) and stomach complaints, also helpful in diabetes. Roots and stem bark are considered expectorant, demulcent and galactogogue. Bark is prescribed for dysentery and biliousness. Dried fruits are useful on intestinal complaints, and also prescribed for colic, dysentery and diarrhoea. Leaves and branches are lopped for fodder. Fibre is obtained from stems which is used for cordage and gunny bags. Wood is used for fuel, charcoal and gun powder.

(Anonymous, 1948–1976)

Ambasta (1986) noted uses similar to those in Anonymous (*loc.cit.*).

Roots are considered as abortifacient, used as anti-galactogogue, they are used as expectorant. Roots are astringent and demulcent. Stem bark is used to treat cholera and diarrhoea. Leaves, flowers and stem bark are used in anaemia. Fruits are use as anthelmenthic, also used in colic, cough, dysentery, fever, flatulence, gastro-intestinal complaints, intestinal germs, to cure rickets of babies, stomach ache and sores. Leaves are applied on eczema. Roots are used as antidote on snake-bite. Flowers are edible. Seed coil is applied to get relief from body pain and labour pain for ladies. Stem bark is used for making various articles like fan. Symbolically twigs are kept on the house or hut of woman in labour pain.

(Jain, 1991)

Critical Notes

Use of fruits noted by the present investigators and those mentioned in the classical literature are more or less similar in nature. It should be assessed on modern scientific lines, if not verified.

**Hemidesmus indicus* R.Br.

Family: Asclepiadaceae, Local Name: Anantmul, Anantwel, Uparsali

Ethnobotanical Notes

Medicine

1. One fist of roots are crushed and kept in water for half an

hour. Half cup of infusion is given on gonorrhoea for seven days.

2. Roots of this plant and paste prepared from stem bark of *Mangifera indica* L., *Moringa olerifera* Lam., *Madhuca longifolia* (Koen) Mac Bride var. *latifolia* (Roxb.) Chevaliar, *Carica papaya* Linn. *Aloe vera* (L.) Burm. *f*., *Dioscorea bulbifera* Linn. and bulbs of *Allium cepa* Linn. It is mixed thoroughly and applied for burning sensation, itching, and rheumatic affection.

3. Root extract is taken orally for about 15 days to increase milk flow.

Notes from Classics

Roots are employed as tonic, demulcent in dyspepsia, skin diseases, syphilis and leucorrhoea. Roots are roasted in plaintain leaves, beaten into a mass with cumin and sugar. It is mixed with ghee and given as a remedy in genito-urinary diseases, also applied for swellings. Roots are useful in chronic rheumatism, cough and diarrhoea. Hot infusion with milk and sugar acts as alterative and tonic in children. Milky juice is applied on conjunctiva and in opthalmia.

(Watt, 1889–1893)

Leaves chewed are said to be refreshing. Latex is used on inflammation of eyes. Root syrup is used as flavouring agent and in preparation of sherbat.

(Anonymous, 1948–1976)

Dried roots are used as demulcent, diaphoretic, diuretic, alterative, prescribed in rheumatism, urinary complaints and skin troubles.

(Ambasta, 1986)

Roots are cooling, used in blood purification, diarrhoea, digestion disorders, dysentery, galactogogoue, person of hypertension, impotency, laziness, menstrual complaints, nervous disorders, skin diseases, stomach ache, spermatorrhoea, tonsillitis, urinal complaints and veneral complaints. Roots are used as antidote on snake-bite and scorpion sting. Roots are edible, used in chutneys and cold drinks. Fruits are used to get relief from cold. Plant is useful on tooth-ache.

(Jain, 1991)

Critical Notes: Use of roots to increase milk flow is additional

****_Hemigraphis latebrosa_ (Heyne ex Roth) Nees**

Family: Acanthaceae, Local Name: Sarambal

Ethnobotanical Notes

Miscellaneous

1. Leaves are cooked and eaten as vegetable.
2. Leaves are also eaten as fodder by goats.

Critical Notes

Value of leaves as vegetable and fodder especially for goats is still unrecorded. The species is being reported useful for the first time.

***_Heracleum grandis_ (Dalz.) S.M.Almeida**

Family: Umbelliferae, Local Name: Bhaphali

Ethnobotanical Notes

Medicine

1. Leaf powder is taken to improve digestion.
2. Entire plant about 15-20 ml thrice a day is given to patients suffering from cholera.

Veterinary Medicine

1. Root extract is given to goats against flatulence.

Miscellaneous

1. Sun dried fruits are used as condiments.

Notes from Classics

Fruits are considered as carminative, stimulant and diuretic.

(Watt, 1889–1893)

Fruits are used as condiments, considered as carminative, stimulant and tonic and are used in form of infusion in gastric and intestinal troubles.

(Anonymous, 1948–1976)

Ambasta (1986) recorded uses similar to those given by Anonymous (*loc.cit.*).

Critical Notes

Uses such as leaves for digestion, of the plant for cholera and roots to treat flatulence in goats are unreported yet.

**Heterophragma quadriloculare* (Roxb.) K.Schum.

Family: Bignoniaceae, Local Name: Waras

Ethnobotanical Notes

Medicine

1. Fresh bark is crushed and this paste is applied on injury caused due to agricultural tools.

Notes from Classics

Thick fluid extracted from wood is applied on skin infections. Wood is used for rough planking, rafters, scanthings, posts, and also used fuel.

(Anonymous, 1948–1976)

Ambasta (1986) mentioned uses as of those given by Anonymous (*loc.cit.*).

Leaves are used on chilblian sores or toes. Wood symbolically repels snake.

(Jain, 1991)

Critical Notes

Fresh bark is helpful to treat injuries. It is an application not recorded earlier.

**Hibiscus aculeatus* Roxb.

Family: Malvaceae, Local Name: Jungli-ambadi

Ethnobotanical Notes

Miscellaneous

1. Root pieces are added in boiling tank of jaggery to settle the contaminations.

2. Fibres are removed from stem and used to tie bundles of fodder.

Notes from Classics

Plant yields fibres.

(Watt, 1889–1893)

Decoction of root bark is given as a remedy for poisons and swellings and for cleaning kidneys. Infusion of roots in water is used as a cooling drink in hot weather. Leaves are cooked and eaten to improve digestion, considered as anthelmintic. Leaf juice mixed with honey in applied in eye diseases.

(Anonymous, 1948–1976)

Infusion of roots is considered as refrigerent. Stem yields fibre suitable for cordage and ropes, but difficult to extract because of prickles. Leaves are eaten after cooking, considered as anthelmintic and improves digestion.

(Ambasta, 1986)

Roots are used in cold drinks. Stem bark yields fibres. Leaves are eaten as vegetable. Seeds yield oil.

(Jain, 1991)

Critical Notes

The technical know-how of the local people to avoid contaminations using roots while preparing jaggery is new and noteworthy.

Hibiscus cannabinus Linn.

Family: Malvaceae, Local Name: Ambadi

Ethnobotanical Notes

Medicine

1. Fibres are obtained from the stem and employed for various purposes.

Miscellaneous

1. Leaves are used as vegetable.
2. Seeds are ground and used as a feed for domestic animals.

Notes from Classics

Flower juice is mixed with sugar and black pepper is a popular remedy for biliousness. Seeds are said to be aphrodisiac and fattening. They are employed externally as poultice for pains and bruises. Leaves are used as pot-herb. Seeds are roasted and eaten. Fibres are extracted from stem stalks. Seed oil forms good lubricant and illuminant.

(Watt, 1889–1893)

Leaves are used as pot-herb. Leaves are considered purgative and aperient. Flower juice mixed with sugar and black pepper is a remedy for biliousness. Seeds are used as cattle and poultry feed. Seed poultice is applied for pains and bruises. Leaves, tops and tender branches are used as cattle fodder. Stem stalks yield fibres used for rope making.

(Anonymous, 1948–1976)

Seeds are considered to be stomachic and aphrodisiac. Seeds yield fatty oil used in soaps, linalium, paints, varnishes, and after refinery for edible purposes. Leaves are used as cattle fodder. Fibres are used as substitue for jute.

(Ambasta, 1986)

Plant is useful on fever. Leaves are considered as purgative. Leaves are eaten as vegetable. Bark yield fibres.

(Jain, 1991)

Critical Notes

Plant species is exotic, native of Tropical America.

**Holarrhena pubescens* (Linn.) Wall.

Family: Apocynaceae, Local Name: Kudai, Dudhkuda

Ethnobotanical Notes

Medicine

1. Decoction of stem bark is useful in stomach-ache. Half a cup of it is advised once a day for 3-4 days.
2. Extract of stem bark is given orally as antidote for snake bite. It results is vomiting and reduces poison.

3. Stem bark is chewed as a remedy against bite of poisonous insect.
4. Follicles are collected after 'Diwali' festival. Latex from them is dropped on rice grains, mixed thoroughly and few grains are asked to chew as a preventive measure against snake-bite in future.

Miscellaneous

1. Bunch of fresh branches are kept in pond for 2-3 days. After softening they are spread in water so that the fishes are diverted towards fishnet.
2. Wood is useful in house or hut constructions.
3. If a leafy twig kept erected in rice fields, it is believed that yield is there by increased. This prevents crop from diseases.
4. On Friday the tree is visited. After worshipping by turmeric powder, Kumkum (red powder) one has to say, "I have helped you, now you will help me" and then return home. On Saturday same tree is selected, stem is cut and a doll is carved out of it. It is wrapped by black cloth, one lemon, nut of 'Biba'. (*Semecarpus anacardium* L. *f*.), shells, (Kawadi) and bangles are tied around neck of doll, then hang the doll in front of door of house or shop. It is believed that the owner leads prosperous and happy life.

Notes from Classics

Bark of stem, roots and seeds are antidysentric and anthelmintic. Seeds are used in acute stage of dysentery when accompanied by fever. A combination of seeds, honey and cotton is used by natives as local application in to the osuteri cases of inflammation of uterus. Fruits are applied in snake bite to allay swelling and irrigation. Leaves are used as fodder in Punjab.

(Watt, 1889–1893)

Stem and root bark are used in the treatment of dysentery, amoebic dysentery and diarrhoea, piles, colic, dyspepsia, chest affection and diuresis, also useful on skin diseases and spleen. Hot decoction of drug is used as a gargle in tooth-ache. Bark powder is rubbed on body in dropsy. Seeds are used in pessaries for promoting conception,for toning vaginal tissues after delivery. Same is used for veterinary cases. Root and leaf powder is administered to stop haemorrhoges after child-birth and also used to stop bleeding from nose. Flowers are eaten and employed on blood diseases. Leaves are used in chronic bronchitis, boils and ulcers. Wood is used for making small articles as combs, toys, spoons, bobbins, cups, knife handles, walking sticks, hookah stem, ploughs and light furniture etc. Wood ash is rich in potash, used as caustic to open abscesses.

(Anonymous, 1948–1976)

Stem bark is astringent, anthelmintic, stomachic, antipyretic, tonic, and anti-dysenteric, used in amoebic dysentery and diarrhoea. Seeds are used for the same purposes as bark. Seed oil is considered as anthelmintic. Wood is used for small articles, as combs, spoons, toys, bobbins, bruises and mathematical instruments.

(Ambasta, 1986)

Root and stem bark are used on rheumatism. Stem bark is used in the treatment of bronchitis, fever, gout, headache, gastric diseases, dysentery, malaria, menstrual complaint, piles, review taste and as stomach-ache. Flowers are used to improve appetite and to cure worms. Seeds are administered on asthma, colic, digestion complaint, leprosy and also used on jaundice. Plant is useful on constipation, epilepsy, labour, menorrhoea, spleen complaint and spermatorrhoea. Stem bark is used as antidote against snake bite. Root bark is used as veterinary medicine. Latex is used to curdle milk. Wood is used for making small articles. Leaves are cooked and eaten as vegetable.

(Jain, 1991)

Critical Notes

Stem bark is reported useful against snake-bite. Its follicle are chewed as preventive measure for snake-bite. This is not documented earlier. Tribals of this district believe in keeping the plant in rice fields which thereby help to increase yield and to prevent from crop diseases. They also have belief in the plants for prosperity of the family.

****Holoptelea integrifolia*** **(Roxb.) Planch**

Family: Ulmaceae, Local Name: Papada

Ethnobotanical Notes

Medicine

1. Stem bark is heated and kept on intestinal torsion.
2. Leaf paste is kept in jaws to get relief from tooth-ache.

Veterinary Medicine

1. Fibres are obtained from stems and tied around the neck of a bullock to cure neck swellings.

Miscellaneous

1. Leaves are crushed and spread in ponds for killing fish.

Notes from Classics

Wood is used for building purposes and cart making.

(Watt, 1889–1893)

Bark may be pulped and made into hardboards and insulation boards. Bark is mucilagenous and used in external applications for rheumatism. Leaves and young fruits are eaten, though not relished by live-stock. Timber is used for small articles like brushes, combs, carvings, furniture, and also used for agricultural implements, plywood, match boxes and paper pulp. Wood is used as fuel.

(Anonymous, 1948–1976)

Bark pulp is made into hard board and insulation boards. Wood is used for light articles, furniture, carts, carriages, bobbins, dugout boats, plywood, cases, splints, match-boxes etc.

(Ambasta, 1986)

Stem bark is applied on bone fracture, rheumatism, scabies, ulcers, and also for ringworms. Leaves are applied on boils, and also used as antidote to snake-bite and scorpion sting. Bark is symbolically tied on arm in hydrocele. Fruits are edible and flowers are eaten as vegetable.

(Jain, 1991)

Critical Notes

Applications of stem bark for intestinal torsion, leaves for toothache and as fish poison, and bark fibres to cure neck swellings are not documented in past.

******Hygrohila shcullii*** (Buch.-Ham.) M.R. & S.M.Almeida**

Family: Acanthaceae, Local Name: Talimkhana, Kaluskanta, Kolusta

Ethnobotanical Notes

Medicine

1. Red lead is sprinkled on the root pieces and 'Dhup' (A frank icense given by censer) is given on burning charcoal and one piece is passed through a ear hole like ring. It cures half head-ache.
2. Root powder is mixed in old stored jaggery in equal proportions. Twenty one tablets are made from this mixture. These are consumed daily for twenty one days against kidney complaints.
3. Decoction of entire plant, about half cup daily for 3 to 4 days, is given to patients suffering from urinal troubles.

Notes from Classics

Roots ashes or entire plant is used as cooling medicine and diuretic in cases of hepatic obstructions, dropsy, rheumatism, and diseases of genito-urinary tracts. Leaves are cooling and diuretic. Seeds are demulcent, diuretic, and as tonic, used in diseases of genito-urinary tracts.

(Watt, 1889–1893)

Roots, leaves and seeds are used as diuretics, and also employed on jaundice, dropsy, rheumatism, anasaraca, and disease of urino-genital tract.

(Anonymous, 1948–1976)

Ambasta (1986) mentioned uses similar to those in Anonymous (*loc.cit.*).

Root paste is applied on swollen face or body. Stem is used on body pain. Leaves are useful in cancer. Seeds are used on jaundice. Entire plant is used on impotency, rheumatism, dropsy, as tonic and diuretic. Tender leaves are used as vegetable in times of scarcity.

(Jain, 1991)

Critical Notes

Applications of roots against kidney complaints and of the entire plants for urinal troubles are yet to be noted elsewhere. Tribals also use roots for head-ache.

**Indigofera cassioides* Rottl. ex DC.

Family: Papilionaceae, Local Name: Unhali

Ethnobotanical Notes

Medicine

1. Green leaves are burnt. The fumes are inhaved by a patient suffering from piles.

Notes from Classics

Decoction of root is given on cough. Powder is applied for chest pains by Santals. Pink flower are sometimes eaten by people of Chotta Nagpur.

(Watt, 1889–1893)

Decoction of leaves and flowers is administered on cough. Flowers are sometimes eaten. Leaves and flowers are lopped for fodder.

(Ambasta, 1986)

Leaves are useful in the treatment of cough. Plant is useful to treat conception, epilepsy, mennorhoea, mucus in urine, labour and swell face in labour. Flowers and pods are edible.

(Jain, 1991)

Critical Notes

Leaves have not been reported earlier useful for piles.

**Indigofera cordifolia* Heyne ex Roth

Family: Papilionaceae, Local Name: Barbad

Ethnobotanical Notes

Miscellaneous

1. Flour of dried seeds is used to prepare bread during famine.

Notes from Classics

Seeds are eaten in times of scarcity and famine.

(Watt, 1889–1893)

Seeds are mixed with bajra or jowar seeds for making bread and consumed.

(Anonymous, 1948–1976)

White seeds resemble poppy seeds, are mixed with bajara or jowar seeds for making bread. Plant is used as cattle fodder.

(Ambasta, 1986)

Leaves are used on mouth swellings.

(Jain, 1991)

**Indigofera glandulosa* Wendl.

Family: Papilionaceae, Local Name: Bhorupdi, Vasolya tarota

Ethnobotanical Notes

Medicine

1. Leaves are burnt and the ash is mixed with coconut oil. It is applied on 'chikhali' an injury caused on foot due to constant touch of mud.
2. Leaf juice is used against eye complaints in elders, harmful for young ones.
3. Fruits are pounded and powder is administered to get relief from stomach-ache.

Notes from Classics

Seeds are employed as nutritive tonic. Seed flour is baked into cake and used as diet during famine period.

(Watt, 1889–1893)

Leaves are used fodder and green manure.

(Anonymous, 1948–1976)

Seeds are considered as nutritive tonic. Seeds are sometimes used as food in scarcity. Plant is used as cattle fodder, may also be used as green manure.

(Ambasta, 1986)

Seeds are edible.

(Jain, 1991)

Critical Notes

Leaves are used for treating 'Chikhali' and eye complaints. Fruits are helpful to control stomach-ache. These are not yet documented so.

**Indigofera tinctora* Linn.

Family: Papilionaceae, Local Name: Diwali

Ethnobotanical Notes

Medicine

1. Fruits are this plant *Cassia tora* L.(Tarota), *Caesalpinia bonduc* (L.) Roxb. (Sagargota) and *Areca catechu* Linn. (Supari) are roasted and crushed. Tablets are prepared and given to patients against flatulence three times per day for 3 to 4 days.

Notes from Classics

Poultice of leaves is applied to perinaeum, which causes flow of urine in spasmodic stricture. Seeds yield oil which is used on sores, promotes growth of hair, externally applied on abdomen in cases of tympanis and retention of urine. Extract of seeds is given as epilepsy, bronchitis and nervous disorders. Seed oil is a remedy for mineral poison and also applied on sores of horses.

(Watt, 1889–1893)

Roots are used on urine complaints. Plant extract is used on epilepsy, nervous disorders and bronchitis. An ointment is applied on sores, ulcers and haemorrhoids. Juice of leaves are used as a remedy for hydrophobia.

(Anonymous, 1948–1976)

Ambasta (1986) gave uses similar as those in Anonymous (*loc.cit.*).

Roots are used to treat fever and to cure wounds. Leaves are used to treat eye diseases, liver diseases and fracture of fingers. Plant is useful on bronchitis, dropsy epilepsy, heart ailments, kidney diseases, nervous disorders, spleen complaints, whooping cough, applied on sores and ulcers.

(Jain, 1991)

Critical Notes

Fruits are reported useful to treat flatulence. It is still unreported.

Ipomoea carnea Jacq.

Family: Convolvulaceae, Local Name: Besharam

Ethnobotanical Notes

Miscellaneous

1. Stem is used for thatching walls of bath room and huts.
2. Dried stems are used as fuel.

Notes from Classics

Tuberous roots and stem are used as food on Hawaii island.

(Ambasta, 1986)

Leaves are used on cuts, wounds, sprain and ulcer. Leaves cooked and eaten as vegetable.

(Jain, 1991)

Critical Notes

The uses although of local importance, the local people have tried to employ this noxious weedy species. It naturally controls its spread to some extent. It is exotic, native of South America.

***Ipomoea hederifolia* L.

Family: Convolvulaceae, Local Name: Dhudhali

Ethnobotanical Notes

Miscellaneous

1. Leaves are cooked and eaten as vegetable.
2. Plant in used as cattle and goat fodder.

Critical Notes

Utility of the leaves as vegetable and the entire plant at fodder are not documented earlier. For the first time, it is being noted as useful species. It is exotic, native Tropical America.

**Ipomoea illustris* (C.B.Cl.) Prain

Family: Convolvulaceae, Local Name: Gogveli

Ethnobotanical Notes

Medicine

1. Leaves are burnt and fumes are passed over the body of children for chicken-pox.

Miscellaneous

1. Stem is splitted and used to tie bundles of fire wood and fodder.

Notes from Classics

Young shoots are eaten in curries. Slender stems are used in cordage.

(Anonymous, 1948–1976)

Slender stems are substituted for cordage. Young shoots are eaten in curries.

(Ambasta, 1986)

Critical Notes

Leaves are beneficial to treat chicken-pox. It is not noted in earlier literature.

**Ipomoea nil* (Linn.) Roth

Family: Convolvulaceae, Local Name: Nilvel

Ethnobotanical Notes

Medicine

1. Leaves are boiled in water and foam is applied on burns for early recovery.

Miscellaneous

1. Plants are used as fodder for goats.

Notes from Classics

Dried seeds are used as purgative. Seeds contain oils which is used as a substitute for Jalap (*Exogonium purga* Benth.).

(Ambasta, 1986)

Seeds are considered as purgative. Entire plant is useful as galactogogue, tonic and also used as vegetable.

(Jain, 1991)

Critical Notes

Applications of leaves for burns and the plants as fodder for goats are not reported yet.

**Ipomoea obscura* (L.) Ker.-Gawl.

Family: Convolvulaceae, Local Name: Pingulvel

Ethnobotanical Notes

Medicine

1. Leaf paste is applied on swellings and knots on body.

Notes from Classics

Leaves and twinning plant have mucilaginous taste, when roasted, powdered and boiled with ghee, considered as valuable application is aphthous affections.

(Watt, 1889–1893)

Anonymous (1948–1976) and Ambasta, (1986) recorded similar uses as those recorded by Watt (*lot.cit.*).

Stem is used in cold and cough.

(Jain, 1991)

Critical Notes

Leaves are valuable in the treatment of body swellings or knots. This is not recorded earlier.

*Ipomoea pes-tigridis L.

Family: Convolvulaceae, Local Name: Bervel

Ethnobotanical Notes

Medicine

1. Decoction of flowers and roots are used on leucorrhoea.
2. Powdered and dried seeds are smoked as a remedy for cough.
3. Flowers, roots and sugar candy (Khadisakhar) are ground together. Tablets are prepared from it. Two tablets are given twice a day for twenty one days to treat impotency.

Notes from Classics

Roots are considered as purgative. Leaves are employed in the form of poultice to boils and sores, also useful on carbuncle and pimples. Plant is used as cattle fodder as green and hay.

(Ambasta, 1986)

Roots are said to be laxative. Roots are used to treat injury of dog-bite. Leaves are used as a remedy for eye complaints, fever and headache. Seeds are edible.

(Jain, 1991)

Critical Notes

Flowers and roots are used against leucorrhoea and impotency. Seeds are applied for cough. These applications have not appeared in the classical literature.

*Ipomoea sepiaria Koen. ex Roxb.

Family: Convolvulaceae, Local Name: Choka Bhaura

Ethnobotanical Notes

Medicine

1. Dried flowers and fine pieces of kernel of coconut are wrapped in the leaves of *Bauhinia malabarica* (Roxb.) Bth. and this bidi is smoked as a remedy on throat infections or inflammation, boils on back and part of tongue.

Notes from Classics

Plant is eaten as pot-herb. Plant juice is considered as diuretic, and deobstruent, also used as antidote to arsenic poisoning.

(Ambasta, 1986)

Root are used in the treatment of diabetes.

(Jain, 1991)

Critical Notes

Use of flowers as noted above is new.

**Ipomoea turbinnata* Lag.

Family: Convolvulaceae, Local Name: Guggalvel, Bhaura

Ethnobotanical Notes

Medicine

1. Dried stem is ground into powder. It is boiled and this decoction is given against cold and fever. About 20 ml of it is advised only once a day till cure.

Miscellaneous

1. Swollen part of pedicel is cooked and eaten as vegetable by tribals.

Notes from Classics

Plant juice is used to destroy bugs. Plant is eaten in times of famine.

(Watt, 1889–1893)

Swollen pedicels are edible. Seeds are cathartic.

(Ambasta, 1986)

Calyx is edible.

(Jain, 1991)

Critical Notes

Stem is not recorded in past literature as useful for cold and fever.

Ixora pavetta Andr.

Family: Rubiaceae, Local Name: Lokhandi

Ethnobotanical Notes

Miscellaneous

1. Stems are used walking sticks.
2. Leaves are used as cattle and goat fodder.
3. Wood is used as fuel.

Notes from Classics

Roots and fruits are useful as medicinally especially to females in case of dark coloured urination. Ripe fruits are eaten by Santals. Leaves are fed to buffaloes. Wood is used for building purposes, and also used as fuel.

(Watt, 1889–1893)

Decoction of bark is used for general debility and anaemia. Ripe fruits are eaten. Leaves are used as fodder for cattle. Wood is used for furniture, building purposes, and engraving and turning articles.

(Ambasta, 1986)

Roots and fruits are used on bleeding at the time of delivery. Entire plant is useful in dysuria, and also used on spider bite poison. Stem bark is used on tooth-ache and work in teeth. Fruits are edible. Wood oil is applied on skin diseases and eczema. Wood is used for basketery and walking sticks.

(Jain, 1991)

Jasminum malabaricum Wight

Family: Oleaceae, Local Name: Kauser

Ethnobotanical Notes

Miscellaneous

1. Fruits are boiled and then eaten as vegetable.
2. Ripe fruits are consumed by tribals.

Notes from Classics

Plant juice is used in the treatment of cataract.

(Ambasta, 1986)

Fruits are edible.

(Jain, 1991)

Critical Notes

The present report increases authenticity of edibility of fruits.

**Jatropha curcas* L.

Family: Euphorbiaceae, Local Name: Moghli-erand, Parsi erand, Chandrajyot

Ethnobotanical Notes

Medicine

1. Stem is used as brush to kill germs in gums.

Veterinary Medicine

1. Crushed fresh twigs are kept in mouth of a cattle against flatulence.
2. Latex is applied on injuries and tumours of cattle.

Miscellaneous

1. Air is passed through a narrow tube into latex and balloons are produced as an amusement by children.

Notes from Classics

Juice extracted from twigs is applied on unhealthy sores and sinuses. Decoction of leaves forms an excellent wash in eczema and indolent ulcers. Seed oil is used as a remedy for itch, herpes, chronic rheumatism and for cleaning application of wounds and ulcers. Seed oil is used for illuminating purposes. Viscid fluid from plant is applied to inflamed gums. Stem sticks are used as tooth brushes against tooth-ache.

(Watt, 1889–1893)

Tender twigs are used for cleaning teeth. Juice from twigs relieves tooth-ache and strengthens gums. Seed oil is externally used for skin diseases and rheumatism. Oil is reported to be abortifacient and efficacious in dropsy, sciatica and paralysis. Seed oil is used on sores of domestic stock. Seeds are considered as purgative. Seed oil is used as illuminant, lubricant, for soap making, candles, and as adulterant in edible oil. Seed cake is used as manure.

(Anonymous, 1948–1976)

Ambasta (1986) recorded uses similar to those given by Anonymous (*loc.cit.*).

Roots are abortifacient. Stem is used for brushing teeth, relieves tooth-ache and tightens gums. Leaves are used on congestion of chest and inflammation. Plant is recommended on anasarca, carbuncle, cramps, pleurisy (lung disease) pneumonia whitlow and syphilis. Whole plant is symbolically wards of evil spirits.

(Jain, 1991)

Critical Notes

Fresh twigs are used as a remedy against flatulence in case of pet animals. This is an additional ethnoveterinary application of this species. It is exotic, native of Tropical America.

**Kedrostis rostrata* (Rott.) Cogn.

Family: Cucurbitaceae, Local Name: Mirchya-kand

Ethnobotanical Notes

Medicine

1. A glass of root extract is given as antidote on snake-bite. This results in vomiting and reduces effect of poison.

Notes from Classics

Root is considered as demulcent, and used in piles and asthma. Fruits and leaves are edible.

(Anonymous, 1948–1976)

Ambasta (1986 reported uses as those in a Anonymous (*loc.cit.*).

Critical Notes

Employment of roots as an antidote for snake-bite is additional.

Kirganellia reticulata (Poir.) Baill.

Family: Euphorbiaceae, Local Name: Kali tahankel

Ethnobotanical Notes

Medicine

1. Oil prepared from roots and stem is applied on rheumatic joints.

Notes from Classics

Decoction of roots is given to children as a remedy for cough, catarrah and asthma. Stem juice is used for sore-eyes in Ghana. Leaves are considered diuretic and cooling. Leaf juice with camphor is given on diarrhoea in case of infants. Leaf powder is used on sores, burns, suppurations and chaffing of skin. Fruits are eaten in times of scarcity. Root is eaten in times of scarcity. Root is used as red dye. Stems are used for hats, baskets and chew-sticks.

(Anonymous, 1948–1976)

Bark is considered to be astringent and diuretic. Fruits are eaten in times of scarcity. Stem is used for making baskets. Bark yield red dye.

(Ambasta, 1986)

Roots are recommended in catarrah and cough. Stem is used as tooth brush. Leaves are considered to be cooling, diuretic and used in diarrhoea and bleeding gums. Leaves are used as fishing bait.

(Jain, 1991)

Critical Notes

Roots are found useful to treat rheumatic joints. This use has not been mentioned in classical literature.

**Kyllinga tenuifolia* Steud.

Family: Cyperaceae, Local Name: Golliband

Ethnobotanical Notes

Miscellaneous

1. Entire plant or piece of plant is kept in 'taite' and tied round the neck for better health and protect from devil's power.

Notes from Classics

Root oil is used for stimulating liver and to relieve prutitus.

(Ambasta, 1986)

Critical Notes

Plant is believed helpful to better health and to protect from devils power. It is an ethnobelief but reported from this district.

**Lagascea mollis* Cav.

Family: Compositae, Local Name: Phulari

Ethnobotanical Notes

Miscellaneous

1. Fresh fruits are edible, consumed by the tribal children.

Notes from Classics

Leaf paste is applied on cuts and wounds. Extract of flower heads, is applied around eyes for conjunctivitis.

(Jain, 1991)

Critical Notes

Edibility of fruits is still unreported.

Lagenaria siceraria Standl.

Family: Cucurbitaceae, Local Name: Dawali, Dawala

Ethnobotanical Notes

Miscellaneous

1. Dried fruits are carved internally and used as water containers. These are used specially when the tribals go out for field work.
2. Fruits are cooked and eaten as vegetable.

Notes from Classics

Roots and seeds are used in the treatment of dropsy. Fruit juice mixed with lime juice is applied on pimples. And if boiled in coconut oil is useful in rheumatism. Fruit pulp is considered emetic and purgative. Tender bottle gourds, shoots and leaves are consumed as vegetable. Seeds are boiled in salt water and eaten as appetizer. Fruit shells are used as water bottles, bowls, ladles, pipes, blowing horns, snuff boxes etc. Dry shells are used for musical instruments like sitar and bin.

(Anonymous, 1948–1976)

Uses recorded by Ambasta (1986) are similar as reported by Anonymous (*loc.cit.*).

Leaves are used on jaundice. Fruits are said to be emetic and purgative, and applied on cracks of sole. Fruits are used as containers like bottles, bowls and utensils.

(Jain, 1991)

Critical Notes

It is a cultivated for its edible fruits which are used as vegetable.

**Lagerstroemia parviflora* Roxb.

Family: Lythraceae, Local Name: Bondara

Ethnobotanical Notes

Veterinary Medicine

1. Leaves are ground and the paste is applied on dislocations, or injury on leg or tail of an animal.

Miscellaneous

1. Wood is employed for hut construction.
2. Wood is used as fuel.
3. A 'Khirla' or 'Ghashya' made up of wood which is pulled by bullocks is used for carrying agricultural implements.

Notes from Classics

Bark yields fibre used for rope making. Bark is used for dyeing skins black. Timber is used for agricultural implements, construction of boats, and also used for charcoal.

(Watt, 1889–1893)

Bark is used for dyeing threads. Leaves are used as cattle fodder. Gum is edible. Timber is used for building purposes, fencing, bridges, boats, oars, agricultural implements, carts, cooperage, water pipes, tanks, boxes, tool handles, tent pegs, wedges, ship's hatches, golf sticks, furniture, etc.

(Anonymous, 1948–1976)

Uses recorded for Ambasta (1986) are similar as reported by Anonymous (*loc. cit.*).

Stem bark is used to increase lactation and to cure wounds. Leaves are used on abscesses. Plant is useful in the treatment of asthma, bronchitis, carbuncle, cough, phthis, sores and syphilis. Stem is used for thatching hut roofs. Wood is used to prepare drums.

(Jain, 1991)

Critical Notes

Additional uses of leaves for dislocations and injury of tail of domestic stocks are being reported.

Lannea coromandelica (Houtt.) Merrill

Family: Anacardiaceae, Local Name: Modhal

Ethnobotanical Notes

Medicine

1. Stem bark is crushed and paste is applied on injury caused by agricultural tools.
2. Stem bark is crushed, heated and applied on swellings.

Miscellaneous

1. Wood is useful for making door frames.
2. Wood is for making drums.

Notes from Classics

Bark is acrid and astringent, is used as lotion for bruises, wounds, sores, ulcers and sore eyes. Decoction of bark is stimulant, used for apthae of mouth and tooth-ache. Leaves are boiled and applied on sprains, bruises, local swellings, body pains and elephantiasis. Juice of twigs mixed with tamarind (*Tamarindus indica* L.) fruits is given as an emetic in cases of coma or insensibility produced by opium or other narcotics.

(Anonymous, 1948–1976)

Bark is said to be astringent. Gum is used as flocculating agent for classification of cane juice. Wood is used locally for house building packing cases, furniture, wheel spokes, cattle yokes, oil presses, rice pounders, ploughs, spears, shafts, combs, brushes, wooden jars, troughs, dugouts and boats, tea-chest, plywood, bobbins, rollers in jute mills, low quality pencils, and also yields paper pulp.

(Ambasta, 1986)

Bark is used on eruptions, stomach-ache, tooth-ache and wounds. Leaves are applied on swelling and ulcers. Plant is used on carbuncle, cholera, diarrhoea, dysentery, haematuria and sores. Wood is used to prepare combs, tobacco pouch etc.

(Jain, 1991)

**Lantana camara* Linn.

Family: Verbenaceae, Local Name: Geltur, Gantur, Gangutai, Ghaneri

Ethnobotanical Notes

Medicine

1. Infusion of roots about two spoons is drunk for seven days, to treat jaundice. It is administered in early morning when stomach is empty.

Miscellaneous

1. Stems are used for thatching hut walls and roofs.
2. Plant is planted on field boundaries for protection of crops.
3. Dried branches are used as fuel.

Notes from Classics

Decoction of roots is used as gargle for tooth-ache. Infusion of leaves is taken internally for bilious fevers and catarrhal affections, externally used as lotion against eczema, eruptions and rheumatism. Leaf oil is used to treat skin diseases and itch, also used as antiseptic on wounds. Entire plant is considered vulnerary, diaphoretic, carminative and antipasmodic. It is useful for fistulae, pustules and tumours. Decoction of plant is given as tetanus, also used in rheumatism, malaria and atoxy of abdominal viscera. Leaves and twigs are used as green manure.

(Anonymous, 1948–1976)

Ambasta (1986) mentioned use similar in Anonymous (*loc.cit.*).

Leaves are used on itching, ring-worms, ulcers, wounds, and at the time of child-birth. Plant is used against malaria, and also useful in rheumatism. Leaves are used as veterinary medicine. Fruits are used as fish bait. Fruits are edible.

(Jain, 1991)

Critical Notes

In the treatment of jaundice roots are being reported beneficial. This is unreported yet. It is exotic, native of Tropical America.

**Launea procumbens* (Roxb.) Ramayya and Rajgopal

Family: Compositae, Local Name: Pathri

Ethnobotanical Notes

Medicine

1. Five to seven leaves are consumed for seven days to reduce piles.

Miscellaneous

1. Leaves are fed to goats.

Notes from Classics

Plant is used in sherbat.

(Watt, 1889–1893)

Leaves are used in curries. Plant is used in sherbat. Plant is used as fodder for goats.

(Anonymous, 1948–1976)

Leaves are used in curries. Plant is used in the preparation of coding sherbat.

(Ambasta, 1986)

Leaves are used on fever. Leaves are eaten as vegetable. Roots are symbolically worm around neck to ward off evil spirit.

(Jain, 1991)

Critical Notes

Medicinal utility of leaves to treat piles is yet not reported.

**Lavandula bipinnatifida* (L.) O.Ktze.

Family: Labiatae, Local Name: Rangajar

Ethnobotanical Notes

Medicine

1. Leaf powder is smoked in bidi prepared using leaf of *Piliostigma malabaricum* (Roxb.) Bth. to cure cough.
2. Seeds are chewed by adults to improve digestion.

Notes from Classics

Root paste or solution is externally applied for scorpion sting or bite of poisonous animals. Tops of plant are eaten by cattle.

(Anonymous, 1948–1976)

Leaves and flowers yield oil.

(Ambasta, 1986)

Plant is used to remove skin outgrowths.

(Jain, 1991)

Critical Notes

Applications of leaves to treat cough and of seeds to improve digestion are unreported yet.

Leea asiatica L.

Family: Vitaceae, Local Name: Gharbandh

Ethnobotanical Notes

Miscellaneous

1. Fruits are sweet and edible. They are consumed by tribals.

Notes from Classics

Root tubers are used for guinea worms. Leaves are applied on wounds. Berries are eaten.

(Anonymous, 1948–1976)

Ambasta (1986) recorded similar uses as given by Anonymous (*loc. cit.*).

Roots are used on rheumatism and urine complaints. Leaves are used on wounds and cuts. Fruits are edible.

(Jain, 1991)

**Leea macrophylla* Roxb. ex Horn.

Family: Vitaceae, Local Name: Sapud

Ethnobotanical Notes

Medicine

1. A glass of root juice is given to reduce poison of snake-bite.

Veterinary Medicine

1. Root ash is diluted in water. About one litre of it is given to horses to cure throat infection.

Miscellaneous

1. Food is served on leaves for the guests on the day of 'Shradha' (offering in memory of the dear departed).

Notes from Classics

Roots are astringent, hence reputed as a remedy for ring-worms and guinea-worms. Pounded roots are applied to obstinate sore and to promote cicatrization. Roots are useful to stop bleeding from wounds. Leaves were used as an article of food in Nasik District during famine on 1877–1878. Leaves are cooked as vegetable and eaten. Large leaves are used as platters.

(Watt, 1889–1893)

Roots tubers are applied to wounds, sores and also useful to treat ring-worms and guinea worms. Leaves are eaten as vegetable. Fruits are edible.

(Anonymous, 1948–1976)

Mucilaginous root tubers are used as anodyne, also useful on wounds, sores, guinea worms and ring worms. Leaves are eaten as a vegetable. Fruits are edible.

(Ambasta, 1986)

Roots are used in the treatment of bone fracture, body pain, sprain, as vermicide, and also used on wounds. Leaves are used on leucorrhoea. Entire plant is useful in neuralgia, chest pain, pleurisy, pneumonia and spleen complaint. Plant is used on veterinary fractures. Leaves are edible as vegetable.

(Jain, 1991)

Critical Notes

Roots are beneficial in case of snake-bite. It is not documented in past.

**Lepidagathis trinervis* Wall. ex Nees

Family: Acanthaceae, Local Name: Gathe-gawat

Ethnobotanical Notes

Medicine

1. Ash of inflorescence is applied to cure swellings and wounds.
2. Infusion of roots is administered to reduce bleeding in piles.

Notes from Classics

Plant is considered as bitter tonic.

(Ambasta, 1986)

Critical Notes

Inflorescence to treat swelling and wounds, and roots for bleeding piles are new applications.

Lepisorus thunbergianus (Kaulf.) Ching

Family: Polypodiaceae, Local Name: Lasani

Ethnobotanical Notes

Miscellaneous

1. Short piece of plant is kept in 'Taite' (amulet) and tied around one's neck to get relief from evil spirit.

Critical Notes

It is an ethobelief.

Leucas cephalotes (Roxb. ex Roth) Spr.

Family: Labiatae, Local Name: Burad

Ethnobotanical Notes

Miscellaneous

1.Plant is used as fodder for domestic animals.

2.Tender leaves are cooked and eaten as vegetable.

Notes from Classics

Plant is a mild stimulant and diaphoretic and is useful on fevers and cough. Leaf juice is externally used on scabies. Flowers are administered on cough and colds. Leaves are used as pot-herb. Flowers are sacred to God Shiva and frequently used as offering.

(Watt, 1889–1893)

Plant is considered as stimulant, diaphoretic, laxative, anthelmintic, antiseptic and insecticidal. A syrup of flower is used as domestic remedy for coughs and colds. Plant is used as pot-herb. Seed oil is used for illumination purposes.

(Anonymous, 1948--1976)

Plant is used as pot-herb. Plant is considered as stimulant, diaphoretic, laxative and anthelmentic. Syrup prepared from flowers is used for cough. Seeds yield oil which is used as a illuminant.

(Ambasta, 1986)

Leaves are used on cold cough, constipation, diarrhoea, fever, headache, piles, scabies, skin diseases, wounds etc. Flowers are used in tooth-ache and gum troubles. Entire plant is used in the treatment of eye diseases, malaria and said to be stimulant and tonic. Plant is recommended as antidote on snake-bite. Leaves are eaten in chutneys and curries.

(Jain, 1991)

Luffa acutangula (Linn.) Roxb.

Family: Cucurbitaceae, Local Name: Dodka

Ethnobotanical Notes

Medicine

1. Leaf paste is applied on body having white spots as a preliminary stage of leprosy.

Miscellaneous

1. Fruits are cooked and consumed as vegetable.

Notes from Classics

Leaves are used as poultice in haemorrhoids, leprosy and splenitis. Leaf juice is reported useful in granular conjunctivitis in children. Decoction of leaves is used on uraemia and amenorrhoea. Plant is useful for skin diseases and asthma, considered as diuretic and given in spleenic enlargements. Fruit powder is used as snuff in jaundice. Seeds are emetic, expectorant and demulcent. Leaves are applied externally to sores of cattle.

(Anonymous, 1948–1976)

Decoction of leaves is used in uraemia and amenorrhoea. Leaves are recommended as poultice in haemorrhoids, leprosy and splenitis. Leaf juice is useful on conjunctivitis.

(Ambasta, 1986)

Fruits are used in fever and jaundice. Plants is used on convulsion, cramps, madness, scabies, syphilis and tetanus.

(Jain, 1991)

****Luffa cylindrica*** **(L.) M.J.Roem.**

Family: Cucurbitaceae, Local Name: Gilke

Ethnobotanical Notes

Medicine

1. Leaves are crushed on hands and paste is rubbed on entire body to reduce fever.

Miscellaneous

1. Fruit sponge obtained from dried fruit is used at the time of bath brushing body and also for cleaning utensils.
2. Tender fruits are cooked and eaten as vegetable.

Notes from Classics

Leaf juice is useful in respiratory complaints. Tender fruits are considered to be diuretic and lactogogue. Ripe fruits after burning and pulverizing are considered as carmanative and anthelmintic. Seed oil is used on skin affections. Loofah sponge is used as bath sponges, cleaning motor cars, glasswares, armoured vehicles, for stuffing pillows, mattresses, saddles, etc. They are preferred as filters, in marine steem engines and used as a substitute for cellulose sponges, Turkish toweling and paper pulp.

(Anonymous, 1948–1976)

Tender fruits are considered as diuretic and lactogogue.

Ambaṣta (1986) recorded uses similar to those in Anonymous (*loc.cit.*).

Fruits are used in diarrhoea, dyspepsia and for prickly heat. Fruits are used as veterinary medicine. Fruits are used for edible purpose.

(Jain, 1991)

Critical Notes

Rubbing of leaf paste on body to reduce fever is a new report to the science. It is a cultigen.

Lygodium flexuosum (L.) Swartz.

Family: Lygodiaceae, Local Name: Haat Zodi

Ethnobotanical Notes

Medicine

1. Rhizomes are rubbed on stone in water and this slurry is applied on injuries. Rhizomes are sold in local markets for Rs.5/- per pair.

Notes from Classics

Plant is used as expectorant. Roots are used in external application for rheumatism, sprains, scabies, eczema, cuts and wounds. Stems may be used for tying rice sheaves.

(Anonymous, 1948–1976)

Ambasta (1986) recorded uses similar as in Anonymous (*loc.cit.*).

Roots are used on abdominal pains, roots are considered as abortifacient, used in antifertility, appetizer, to treat cholera, jaundice, stomach-ache, improves indigestion, debility and applied on cuts, eczema, ulcers, scabies and wounds. Plant is used on constipation, cough and pleurisy. Frond is used on fever and rheumatism. Fibre is used in construction work and as cordage. Symbolically occurrence in agricultural fields is auspicious, foretells poor crop.

(Jain, 1991)

Critical Notes

Rhizome is applied to cures injuries. This application is new. In tribal areas, it is sold.

Madhuca longifolia (Koen.) MacBride var.*latifolia* (Roxb.) Chevalier

Family: Sapotaceae, Local Name: Mahu, Mahua

Ethnobotanical Notes

Miscellaneous

1. Stem pieces are made into ring and tied with bamboo strips to make "Tondya" used for fish trapping.

2. Leaves are used for thatching roofs of huts.
3. Flower buds are used for preparing wine.
4. Seeds are crushed and oil is extracted, used for edible purposes and burning lamps.
5. Seeds oil is given to bullocks to improve digestion.
6. Gum is used to avoid disturbance of birds in crop fields. Gum is applied on a stick or log called "Chikadi". It is kept in crop fields.
7. Latex is applied on arrows and kept stored in a bamboo container. They are used to kill birds.

Notes from Classics

Decoction of bark is astrigent and tonic, used as a remedy for rheumatic affections. Decoction of flowers is used in coughs. Plant attributes properties as stimulant, demulcent and emollient, heating, astringent, tonic and nutritive. Flowers mixed with milk are used in impotence due to debility. Corollas are eaten raw or cooked. A spirituous liquor is extracted from flowers. Fruits are some times eaten. Seed oil is recommended to the head in cephalgia, also used on skin diseases. Seed cake is employed as emetic and as a detergent. Residue of seed after expressing oil is used as fish poison. Wood is used for furniture, doors, window, panels, naves of wheels etc. Leaves are used as platters.

(Watt, 1889–1893)

Roots are applied on ulcers. Flowers are regarded as demulcent, used in cough, cold and bronchitis. Honey from flowers is used for eye diseases. Seeds are used as galactogogue. Seed oil is used on skin diseases, rheumatism and head-ache. Oil is considered as laxative, useful in habitual constipation, piles and haemorrhoids, also used as emetic. Milky latex is used on rheumatism, ulcers, itches, bleeding and spongy gums, tonsillitis and diabetes. Mahua berries are eaten raw or cooked. Seed is mainly used in manufacture of soaps, lubricating greases and fatty alcohols, candles, and as raw material is production of stearic acids. Seeds cake is fed to animals, also used as manure. Wood is used for building purposes, carriages, turnery, sports goods, musical instruments, bridges, agricultural implements, carvings and also used as fuel wood. Leaves are eaten by cattle. Latex is given to horses in stomach-ache.

(Anonymous, 1948–1976)

Uses recorded by Ambasta (1986) are similar as given by Anonymous (*loc. cit.*).

Stem bark is astringent, used in the treatment of diabetes, gum trouble, itch and rheumatism, also used as tonic. Leaves are used on burns and scalds. Flowers are used on bronchitis, cough and cold. Seed oil is applied on chest pain and pneumonia. Stem bark or flowers are used as antidote against snake bite. Seed oil is used for luminating lamps. Wood is used miscellaneous purposes.

(Jain, 1991)

Critical Notes

Local uses of leaves, seed oil, gum and latex are new and noteworthy.

Mallotus philippensis Muell.-Arg.

Family: Euphorbiaceae, Local Name: Kukwali

Ethnobotanical Notes

Miscellaneous

1. Powder obtained from fruits is used as 'Kumkum' by tribals.

Notes from Classics

Leaves, fruits and bark are considered to be anthelmintic. Kamela powder is anthelmintic, vermifuge, and cathrtic, externally used in scabies and skin diseases. Seed oil is used in anorexia, fever, giddiness, hemiplegia, hepatic and thorasic pains.

(Watt, 1889–1893)

All parts of tree are used in external applications for skin diseases. Leaves are used as cattle fodder. Wood is used for rafters, tool handles, match-boxes, small turnery articles, paper pulp and is also used as fuel wood.

(Anonymous, 1948–1976)

Glandular hairs yield kamela powder, considered as anthelmintic, used for cutaneous affections. Seed oil is used in paints, varnishes, hair fixers and ointments. Seed cake in combination with saw dust is used for making insulating boards and cork substitutes.

Seed cake is also used as manure. Wood is used for rafters, tool handles, match-boxes and turnery articles. Leaves are used as cattle fodder.

(Ambasta, 1986)

Roots are considered as tonic for pregnant women and also applied for skin diseases. Fruits are used on blisters, boils, constipation, contraception, dysentery, leprosy, skin diseases and also used as vermifuge, vermillion and on intestinal worms. Seeds are considered as anthelmintic, purgative, used on rheumatism and ulcers. Seeds are treated against snake-bite. Seed oil is used on varnishes. Fruit dye is used for textile dyeing.

(Jain, 1991)

**Mangifera indica* Linn.

Family: Anacardiaceae, Local Name: Amba

Ethnobotanical Notes

Medicine

1. Stem bark is crushed in water. This extract mixed with honey in 1:1 ratio is given on jaundice. Two to three spoons are administered at morning till cure.
2. Bark infusion, about half cup, is given twice a day for relief in asthma.

Veterinary Medicine

1. Unripe fruits are crushed in water and infusion is given to chicks to control dysentery.

Miscellaneous

1. Pickles are made from unriped fruits, and ripe fruits eaten raw.
2. Wreaths of leaves are used in various ceremonies to decorate Pandals, or places of worships, temples etc.
3. Wood is used for agricultural tools.
4. Wood is used as fuel.

Notes from Classics

Leaves are used as platters, also used in adorning mandap and houses on various ceremonies. Twigs are used as tooth brushes. Flowers are used to worship God Shiva on the day of Shivratri in month of Magh. Wood is used for planking doors, windows for packing cases etc.

(Watt, 1889–1893)

Bark is used on diptheria and rheumatism. Fumes from the burning leaves are inhaled for relief from hiccups and affection of throat. Unripe fruits are usually used in pickles, chutneys, amchur and culinary purposes, also used in manufacture of jam, jellies, murrabba, squash, juice etc. Ripe and unripe fruits are eaten. Dried flowers are given on diarrhoea, chronic dysentery, catarrah of bladder and as galactogogue. Ripe and unripe fruits are eaten. Gum is used on cracks of feet and scabies, also recommended in syphilis. Kernel powder is used in bleeding piles. Roasted or boiled kernels are used as human food in times of scarcity. Kernel flour is soaked in water and consumed. Dried kernels are used as feed for poultry. Timber is used for making furniture, house construction, agricultural implements, match boxes, Boats and dugouts, plywood, shoe heels and also used for high value charcoal.

(Anonymous, 1948–1976)

Uses recorded by Ambasta (1986) are similar as given by Anonymous (*loc.cit.*).

Stem bark is used in antifertility, jaundice and stomachache. Fruits are used in diarrhoea and also improves digestion. Riped fruit pulp is laxative, also used as tonic. Gum is used in tooth-ache. Latex is used for application of scabies and styes. Wood is used for making drums and other musical instruments. Tree is symbolically abode of God, leaves are used to decorate on various festival days.

(Jain, 1991)

Critical Notes

Stem bark is used for asthma, whereas the unripe fruits are beneficial to control dysentery in fowls. Both applications are new.

**Marsilea minuta* L.

Family: Marsiliaceae, Local Name: Zar-zurya

Ethnobotanical Notes

Medicine

1. Leaves are boiled slightly and then eaten as vegetable by anaemic persons.
2. Extract of leaves is applied to cure swelling on inflamed spots on face, around eyes or stomach.

Notes from Classics

Plant is used as pot-herb.

(Watt, 1889–1893)

Stalks and leaves are eaten as pot-herb in times of scarcity.

(Anonymous, 1948–1976)

Plant is used as pot-herb.

(Ambasta, 1986)

Plant is used in the treatment of bone fracture, also used as remedy for eye complaints, fever, muscle pain, menorrhoea, pimples and spleen complaints. Stem and leaves are used as vegetable.

(Jain, 1991)

Critical Notes

Plant is widely documented as edible. Presently it is noted especially for anaemic persons, which is not so documented earlier.

Maytenus senegalensis (Lamk.) Exell.

Family: Celastraceae, Local Name: Henkal

Ethnobotanical Notes

Medicine

1. Leaves are chewed to get relief from cough, and also used as mouth wash.

Miscellaneous

1. Leaves are fed to goats.
2. Dried branches are used for fencing farmyards and cultivated fields.

Notes from Classics

Bark paste with mustard oil is applied on head to destroy pediculi.

(Watt, 1889–1893)

Root is used in gastro-intestinal troubles, especially dysentery. Bark paste with mustard oil is applied on hair to kill lice. Pulverized leaves are given with milk to children as vermifuge. Decoction of leafy twigs is used as mouth-wash to relieve tooth-ache. Leaf ash and ghee is applied on sores as an ointment. Ashes of wood and leaves are used as substitute for salt in Nigeria. Leaves are used as fodder. Branches are employed as dunnage for roofs of houses.

(Anonymous, 1948–1976)

Decoction of shoots is used for colic, dysentery and diarrhoea among children. Bark paste with mustard oil is used to destroy pediculi. Leaves are used as fodder. Branches are used for roofing houses.

(Ambasta, 1986)

Leaves are used on eye complaints. Fruits are used for blood purification. Roots are used as veterinary medicine for rejuvenator. Fruits are edible. Twigs are used in various religious ceremonies.

(Jain, 1991)

Critical Notes

Leaves are being reported useful for cough for the first time.

**Melia composita* Willd.

Family: Meliaceae, Local Name: Nimbara

Ethnobotanical Notes

Medicine

1. Stem bark is crushed and its extract about one spoon is given to children as a remedy for intestinal worms. It is given only for three days with milk or water after does.

2. Bark pieces are heated and mounted on stomach of children against stomach worms.
3. Stem bark is chewed orally against snake bite reduces poison.

Miscellaneous

1. Wood is used for preparing 'Ghashya' as carrier pulled by bulls used for carrying agricultural needs.

Notes from Classics

Fruit pulp is used as remedy for labouring classes for colic. Wood is used to prepare boats and tea-boxes.

(Watt, 1889–1893)

Fruits are bitter and considered anthelmintic. Wood is used for building purposes, agricultural implements, packing cases, cigar boxes, pencils, match- boxes, splints and catamarans.

(Anonymous, 1948–1976)

Fruits is anthelmintic, useful in skin infections. Wood is used for packing cases, cigar boxes, ceilings planks, agricultural implements, match boxes, pencils, and catamarans, also suitable for musical instruments, tea chests, and plywood making.

(Ambasta, 1986)

Uses recorded by Jain (1991) are similar to those of Ambasta (*loc.cit.*).

(Jain, 1991)

Critical Notes

Employing stem bark against intestinal or stomach worms is a new application.

**Memecylon umbellatum* Burm.*f.*

Family: Melastomaceae, Local Name: Karap

Ethnobotanical Notes

Miscellaneous

1. Wood used for cart making, specially axis and poles of carts.

Notes from Classics

Bark is applied on bruises. Leaves are supposed to be cooling and astringent, given internally in gonorrhoea and leucorrhoea. Lotion of leaves is applied on conjunctivitis. Ripe berries are eaten by natives.

(Watt, 1889–1893)

Anonymous (1948–1976) mentioned uses similar to those in Watt (*loc.cit.*), except of wood used for tool handles, combs, walking sticks and decorative works.

Leaves are astringent. Their lotion is used for eye complaints. Fruits are edible. Wood is used for rafters, tool handles, pestles, walking sticks, decorative work, also used good quality charcoal.

(Ambasta, 1986)

Flowers and leaves yield yellow dye.

(Jain, 1991)

Critical Notes

Employment of wood particularly for making bullock carts is new.

**Meyna laxiflora* Roybns

Family: Rubiaceae, Local Name: Aliv

Ethnobotanical Notes

Medicine

1. Fine pieces of stem bark are sprinkled by red lead and fumed 'Dhup' by on a burning charcoal and bark piece is tied on the opposite ear to eye having catarrah.
2. Coconut oil is applied on leaf of this plant and slightly heated on fire. It is applied on swellings on inflamed parts.
3. Seed powder is administered twice a day for 8–10 days to cure kidney stones.

Miscellaneous

1. Ripe fruits are eaten by natives.

2. Slender stems are used as walking sticks while circumambulation Bramhagiri mountain at Trimbakeshwar.

Notes from Classics

Dry fruits are believed to be narcotic and is reported useful on boils and dysentery. Leaf powder is used on diptheria.

(Anonymous, 1948–1976)

Leaves afford poor quality fodder. Powdered leaves are useful in diptheria. Dry fruits are having narcotic properties, used in dysentery and boils. Fruits are edible.

(Ambasta, 1986)

Leaves are used in colic pains and against dysentery. Seeds are used in diptheria. Plant used on fissures in skin. Fruits are edible.

(Jain, 1991)

Critical Notes

Uses of stem bark for catarrah and of seeds for kidney-stone are not noted earlier. Leaves are likewise being noted for the first time for treating swellings or inflamed parts.

**Miliusa tomentosa* (Roxb.) J.Sinclair

Family: Annonaceae, Local Name: Humb.

Ethnobotanical Notes

Medicine

1. Unripe fruits are burnt and the smoke is passed over the body of a lady after delivery for body swellings.

Miscellaneous

1. Ripe fruits are tasty and consumed by tribals.

Notes from Classics

Fruits are edible. Leaves are used as cattle fodder. Wood is used for building huts and sheds.

(Watt, 1889–1893)

Leaves are lopped for fodder. Fruits are eaten. Gum is used as substitute for tragacanth. Wood is used building huts, cattle sheds, cabinet works, carving, slack cooperage, bobbins, billiard cues and as fuel wood.

(Anonymous, 1948–1976)

Leaves are used as cattle fodder. Fruits are edible. Wood is used for building huts, cabinet work, slack co-operage, bobbins, shutters, billiards cues, and also used as fuel.

(Ambasta, 1986)

Bark is used on paralysis.

(Jain, 1991)

Critical Notes

Edibility of fruits is also noted earlier, however, their application to reduce swelling especially after delivery is still not documented.

**Mimosa hamata* Willd.

Family: Mimosaceae, Local Name: Aarti, Arkade

Ethnobotanical Notes

Miscellaneous

1. Dried branches are used for hedge.
2. Leaves are fed to goats.
3. Wood is used to make handles of agricultural tools.

Notes from Classics

Plant is used as cattle fodder and also used as green manure.

(Ambasta, 1986)

Leaves are used on chronic cough and pharyngitis. Seeds are used for blood purification.

(Jain, 1991)

Critical Notes

Use of the plant species as hedge and for tool handle are not reported in past.

Mitragyna parvifolia (Roxb.) Korth.

Family: Rubiaceae, Local Name: Kalamb

Ethnobotanical Notes

Miscellaneous

1. Stem bark is chewed and rubbed on body before bath after returning from funeral. It is thought that one's body is thereby purified.
2. Leaves are used as fodder for cattle.
3. Wood is employed for hut and house construction.

Notes from Classics

Bark and root is given in fevers and colic. Bark paste is applied for muscular pain. Leaves are used as fodder. Timber is used for furniture, planking, agricultural implements and various other articles.

(Watt, 1889–1893)

Uses recorded by Anonymous (1946-76) and Ambasta (1986) are similar as noted by Watt (*loc.cit.*).

Stem bark is used on fever, contraception, muscle pain, stomach-ache, syphilis, and leucorrhoea. Leaves are used on head-ache, also useful to treat wounds. Plant is used in carbuncle, cholera, night blindness, odema, sores and on joint pains.

(Jain, 1991)

Critical Notes

Stem bark finds place in funeral rites.

**Momordica dioica* Roxb. ex Willd.

Family: Cucurbitaceae, Local Name: Karle

Ethnobotanical Notes

Medicine

1. Few drops of leaf juice are poured is nostrils to get relief from half head-ache.

Miscellaneous

1. Root tubers are rubbed and the slurry obtained is given on snake bite.
2. Fruits are salted and cooked, consumed as vegetable.

Notes from Classics

Mucilagenous roots are used in cases of bleeding piles and certain bowel affections. Root juice is used as domestic remedy for inflammation caused by contact with urine of house lizard. Powder of dried fruits is used as snuff to result in repeated sneezing. Tubers of female plants are considered as expectorant. Roots of male plant is used on ulcers caused due to snake bite. Unripe fruits are used as vegetable to control fever. Tender and green fruits are eaten by natives.

(Watt, 1889–1893)

Roots are applied on bleeding piles, bowel affection and urinary complaint. Root paste is applied as sedative in fever. Fruits are used as vegetable. Oil obtained from seed kernels is used in paints and varnishes.

(Anonymous, 1948–1976)

Uses notes by Ambasta (1986) are similar as recorded by Anonymous (*loc.cit.*).

Roots are used on infertility, to cure fever and piles. Fruits are used for pain in breast due to swelling. Fruits are edible, used in the treatment of asthma, pithis, post-natal complaints, delirium, dysentery, spleen complaints, head-ache and also used in as veterinary medicine.

(Jain, 1991)

Critical Notes

Applications of leaf juice for head-ache and root tubers against snake-bite are new reports

Morinda pubescens J.E.Sm.

Family: Rubiaceae, Local Name: Ali

Ethnobotanical Notes

Miscellaneous

1. Ripe fruits are consumed by tribals.

2. Wood is used for huts and sheds, also used for agricultural tools.

Notes from Classics

Roots are used internally as an astringent. Green fruits are pickled and eaten with curries.

(Watt, 1889–1893)

Bark yields 'Al' dye. Fruits are eaten. Wood is used for making plated and dishes.

(Anonymous, 1948–1976)

Roots are source of 'Al' dye. Wood is used for making plates and dishes. Fruits are eaten.

(Ambasta, 1986)

Roots are used on ulcers. Bark is used on urine complaint. Fruits are used on bird-lime. Timber is used for miscellaneous articles.

(Jain, 1991)

Moringa cocanensis Nimmo ex Dalz. & Gibs.

Family: Moringaceae, Local Name: Ranshegu

Ethnobotanical Notes

Medicine

1. Stem bark is roasted and tied on leg for pain in one leg.
2. Extract of stem bark about 15-20 ml is drunk to cure stomach-ache.
3. Stem bark is heated and kept on stomach of a pregnant women for abortion.

Notes from Classics

Flowers and fruits are eaten as pot-herb. Roots are used as substitute for horse-radish.

(Watt, 1889–1893)

Young bark and roots are rubefacient and vesicant. Leaves are used as scurvy and catarrhal affections, also used as emetic. Leaf paste is used on wounds. Flowers are used as tonic and diuretic. Seeds are used as tonic and diuretic. Seeds are antipyretic, acrid and bitter. Seed oil is applied on rheumatism and gout. All parts of tree are considered as medicinal,used in the treatment of ascites, rheumatism and as cardiac and circulatory stimulants. Roots are used as condiments. Tender pods are esteemed as vegetable. Seeds are consumed after frying. Twigs and leaves are lopped for fodder. Plant is useful on venomous bite.

(Anonymous, 1948–1976)

Ambasta (1986) recorded similar uses as those given by Anonymous (*loc.cit.*) except the following.

Seed oil is used in cosmetics and illuminant. Seed cake is used as fertilizer. Gum is used in calico printing.

(Ambasta, 1986)

Roots are used as tonic for giddiness. Seeds are used in safe throat.

(Jain, 1991)

Critical Notes

Stem bark is employed for pains in legs, stomach-ache and abortion. All uses are still untapped.

**Moringa oleifera* Lam.

Family: Moringaceae, Local Name: Shevga

Ethnobotanical Notes

Medicine

1. About 20-25 ml of decoction of leaves is given once a day to treat tooth-ache.

Miscellaneous

1. Pods are cooked and eaten as vegetable.

Notes from Classics

Roots are recommended for soreness of mouth and throat. Root is described as acrid, pungent, stimulant, diuretic when given internally and rubefacient when applied externally. Decoction of root bark is recommended for ascites, enlargement of spleen or liver, externally applied over inflamed parts. Flower expel cold humours, disperse swellings, act as tonic and diuretic. These increase flow of bile. Pods are eaten as vegetable, acts preventive against intestinal worms. Seed oil relieves pain of joints, in gout and acute rheumatism. Fruit is administered in affections of liver and spleen, articular pain, tetanus, debility of nerves, paralysis, pustules and leprosy. Leaf juice is given as emetic. Gum is used to produce abortion. Leaves, flowers, pods and twigs are cooked in many ways as pot-herb. Twigs and leaves are lopped as fodder.

(Watt, 1889–1893)

Roots are rubefacient and vesicant. Leaves are used in scurvy and catarrhal affection, also used as emetic. Flowers and tender leaves are eaten as pot herb, considered as tonic, diuretic, and cholagogue. Seeds are fried and consumed. Seed oil is used in rheumatism, also used in cosmetics, and as illuminant purposes. Seed cake is used as fertilizer. Gum is used in calico printing. Twigs are lopped for fodder. All parts of three are useful in the treatment of ascites, rheumatism, as cardiac, and circulatory stimulants, also used against venomous bite.

(Anonymous, 1948–1976)

Ambasta (1986) recorded similar uses these recorded by Anonymous (*loc.cit.*).

Roots are used on asthma, epilepsy, liver complaints, spleen enlargement, paralysis and rheumatism. Bark and gum are abortifacient. Gum is used in ear complaint and head-ache. Bark is used on baldness, carries, cholery, dysentery, eczema, menstrual complaint, stomach-ache and also useful in typhoid. Leaves are used on scurvy and wounds. Fruits are used to improve appetitie, to treat intestinal worms, used as tonic, cooked and eaten or pickled. Flowers are stimulant and eaten as vegetable. Entire plant is useful on antifertility, dysentery, madness, pneumonia, scabies, scrofula, syphilis and urine complaint. Plant is used against scorpion bite. Leaves and twigs are used as fodder for cattle.

(Jain, 1991)

Critical Notes

Utility of leaf decoction to treat tooth ache forms the first report for science.

**Mucuna prurita* Hook.

Family: Papilionaceae, Local Name: Khajol, Kachquiri

Ethnobotanical Notes

Medicine

1. Prickles removed from fruits are mixed in jaggery and homogenized thoroughly. It is given to children against stomach worms.

Miscellaneous

1. Fruits are laid in foundation of new houses by Marwadis. It is believed helpful to lead prosperous life.
2. Tender leaves are cooked as vegetable.

Notes from Classics

Infusion of roots with honey is given in cases of Cholera morbus. Root paste is applied on body in dropsy or root piece in tied around the wrist or ankle as a remedy. Seeds of same plant and fruits of *Tribulus terestis* L. with sugar and cow milk are pounded together and used as aphrodisiac. Roots are tonic used in facial paralysis and hemiplegia. Seeds are applied on part sting by scorpion to absorb poison. Seed powder is given in leucorrhoea. Young tender pods are cooked and eaten as vegetable.

(Watt, 1889–1893)

Roots are considered as tonic, stimulant, diuretic, purgative and emmenogogue, used for diseases of nervous system, kidney troubles and dropsy. Root ointment is used on elephantiasis. Leaves are useful on ulcers. Infusion of hairs is used on liver diseases and gall bladder, applied externally as stimulant and mild vesicant. Bristly hairs are formerly used as vermifuge. Leaves and pods are used as vegetable. Plant is useful as fodder and green manure.

(Anonymous, 1948–1976)

Uses noted by Ambasta (1986) are similar as noted by Anonymous (*loc. cit.*).

Roots are used on cough and dysentery. Stem is used on eye diseases. Leaves are used on eye diseases. Leaves are used on blisters, boils and ulcers. Fruits are considered as anthelmintic. Fruits and hairs are used to treat round worms, and also used as vermifuge. Seeds are useful in appetite, improves digestion, also used in the treatment of impotency and night dreams. Roots are used as veterinary medicine. Seeds are useful against scorpion bite. Fruits are cooked and eaten as vegetable.

(Jain, 1991)

Critical Notes

Application of prickles from fruits is beneficial against stomach worms. It is a new medicinal report.

**Mukia maderaspatana* (L.) Roem.

Family: Cucurbitaceae, Local Name: Mekka, Golmekh.

Ethnobotanical Notes

Medicine

1. Ash from burnt leaves is mixed in coconut oil and applied on foot and between fingers in case of 'Chikhali' or 'Golasti', an injury caused due to constant contact of mud.

Miscellaneous

1. Ripe fruits are eaten.

Notes from Classics

Roots when masticated relieves tooth-ache. Lactation of root is given in flatulence. Plant is given in flatulence. Plant is sometimes prescribed on cough. Tender shoots and bitter leaves are used in vertigo and biliousness.

(Anonymous, 1948–1976)

Uses noted by Ambasta (1986) are similar as recorded by Anonymous (*loc. cit.*).

Roots are used on tooth-ache. Seeds are used as remedy on diarrhoea. Fruits are edible.

(Jain, 1991)

Critical Notes

Leaves are being reported useful for the first time especially on injuries in between foot fingers caused due to constant contact of mud or muddy water.

**Musa paradisiaca* Linn.

Family: Musaceae, Local Name: Keli

Ethnobotanical Notes

Medicine

1. Root pieces are kept in hair for easy delivery.

Miscellaneous

1. Leaves are used as platters.
2. Suckers are used for worship in 'Satayanarayana' and to decorate chariots at the time of pilgrims.
3. Unripe fruits are cooked as vegetable.

Notes from Classics

Roots and stem are considered as tonic, antiscorbutic and used in disorders of blood and veneral diseases. Tender root juice is used to check haemorrhages from the genital and air passages. Juice of flowers mixed with curd is used in dysentery and menorrhagia. Unripe fruit powder is used in gonorrhoea. Fruits are eaten raw or cooked. Leaves are used as platters, and also for packing purposes.

(Watt, 1889–1893)

Ash of roots is anthelmintic. Juice of root and stem is used in disorders of blood. Ripe fruits are used in diabetes, uremia, nephritis, gout, hypertension and cardiac diseases. Fruits are useful in diarrhoea, dysentery and promote healing of intestinal lesions in ulcerative colitis. Banana powder is used in the treatment of celiac diseases, sprue and carbohydrate intolerance in children. It is also useful in intestinal disorders of adults. Banana flour is used as infant food.

(Anonymous, 1948–1976).

Banana fruits are laxative, used in instestinal disorders, uraemia, nephritis, hypertension and other vascular diseases. Pseudostems yield starch used for finishing of textiles. Banana fibre is used for cordage, mats, and for making coarse paper. Fruit flour is used for jams, jellies and confectionary. Unripe fruits are eaten as vegetable. Fruit peels are fed to cattle and goats.

(Ambasta, 1986)

Leaves and flowers are used on asthma, whooping cougn and boils. Plant is used on ansarca, diarrhoea, haematuria, dropsy, mucus in urine, small pox, and vomiting. Flowers, roots and psuedostem are useful in the treatment of dysentery and also useful as antifertility. Flowers and fruits are edible. Stem is used for decorations in various ceremonies.

(Jain, 1991)

Critical Notes

Roots are thought useful for easy delivery, a use hitherto unreported.

*******Nervilia plicata* **(Andr.) Schlecht.**

Family: Orchidaceae, Local Name: Dukar Kand, Dudri

Ethnobotanical Notes

Veterinary Medicine

1. Bulbs are fed to milking cows and buffaloes to increase lactation.

Critical Notes

The plant is being recorded as useful for the first time.

******Nicandra physalodes* **(L.) Gaertn.**

Family: Solanaceae, Local Name: Dhodana

Ethnobotanical Notes

Miscellaneous

1. Ripe fruits are consumed by tribals.

Notes from Classics

Plant is said to be diuretic.

(Watt, 1889–1893)

Plant posses diuretic, anthelmintic and insecticidal properties. It is said to be used as fly poison. Decoction of leaves is used for killing head lice.

(Anonymous, 1948–1976)

Plant is considered as diuretic, insecticidal and anthelmintic, used as fly poison. Seed oil is used in varnishes.

(Ambasta, 1986)

Critical Notes

Edibility of fruits is not reported in past literature. It is exotic, native of Peru. It has naturalized well in the district.

**Ocimum basilicum* Linn.

Family: Labiatae, Local Name: Ran-tulas, Sabja

Ethnobotanical Notes

Medicine

1. Leaves are chewed as a treatment for stomach worms in children.
2. Leaf juice is dropped in ear as a remedy for ear-ache.

Miscellaneous

1. Plant is kept in house as a repellent for mosquitoes and flies.

Notes from Classics

Roots are used on bowel complaints of children. Leaf juice is dropped in ears to cure ear-ache and dullness of hearing. Leaf juice is useful to cure ring-worms. Bruised leaves are applied on part of scorpion sting. Leaf powder is useful in dislodging of maggots. Seed poultices are used for unhealthy sores and sinuses. Seeds are given with sherbet in cases of habitual constipation and in internal piles, also useful for fever. Seeds are demulcent, stimulant, diuretic and diaphoretic. Infusion of seeds is used on gonorrhoea, nephritic affections, dysentery and diarrhoea.

(Watt, 1889–1893)

Seeds posses demulcent and cooling properties, given internally in case of habitual constipation of piles. They are also used as poultices for sores and sinuses. Infusion of plant is useful in cephalgia and gouty joints, used as gargle for foul breadth, also useful as nasal douche and as a nostrum for ear-ache and ringworms. Plant oil possesses insecticidal and insect repellent properties, effective against houseflies and mosquitoes.

(Anonymous, 1948–1976)

Leaf juice is used as a nasal douche for ring-worms and ear-ache. Seeds are demulcent, stimulant, diaphoretic, used in cases of habitual constipation and piles. Poultices of leaves is used for sores and sinuses. Plant is considered to be stomachic, antipyretic, diaphoretic, expectorant, carminative and anthelmintic. Plant yields volatile oil used in perfurmery, confectionary, baked goods, sauces, pickles, vinegar, spiced meat, beverages, and as insecticidal and mosquito repellant.

(Ambasta, 1986)

Roots are used in the treatment of bowel complaint. Leaves are useful to treat cholera, cough, dysentery, stomach-ache, also removes alcohol in toxic. Inflorescence paste is applied on wounds. Seeds are given on diarrhoea. Plant is used on anasarca, cramp, dropsy, ear-ache, epilepsy, fever and labour complaint. Plant is used as antidote on snake-bite.

(Jain, 1991)

Critical Notes

To treat stomach worms, leaves are not reported useful earlier.

Ocimum gratissimum Linn.

Family: Labiatae, Local Name: Ran tulas

Ethnobotanical Notes

Medicine

1. Leaves are added in bath water which is used especially to treat body ache.
2. Leaves are kept between jaws as a remedy for tooth-ache.
3. Plant is used as mosquito-repellent.

Notes from Classics

Leaf juice is a remedy for gonorrhoea. Herb is used in bath and fumigation is given for rheumatism and paralysis. Decoction of leaves is used in case of seminal weakness. Seeds are given in head-ache and neuralgia.

(Watt, 1889–1893)

Leaf juice is given on stomach-ache. Seeds are administered on head-ache, neuralgia and dysentery. Infusion of seeds is used on urinary disorders. Plant yields oil which is used for inflamed joints, at chronic condition and also gives relief in ear-ache, tooth-ache and abdominal colic in children. Plant is used as mosquito and pest repellent.

(Anonymous, 1948–1976)

Seeds are given on head-ache, neuralgia and dysentery. Infusion is used in urinary disorders. Decoction is used in seminal weakness. Plant is considered as digestive tonic, stimulant, demulcent, diuretic, anthelmintic, antiseptic and styptic, employed in cough mixtures. Herb is also useful in bath and fumigations, is given for rheumatism and paralysis. Plant is used against tooth-ache, ear-ache and in abdominal colic in children. Plant is used as mosquito repellent.

(Ambasta, 1986)

Leaves are useful in intestinal diseases and as a remedy for gonorrhoea.

(Jain, 1991)

Critical Notes

Leaves are beneficial in treating body-ache and tooth-ache.

Ocimum tenuiflorum L.

Family: Labiatae, Local Name: Tulsi

Ethnobotanical Notes

Medicine

1. Leaf juice is given to babies to cure cough.
2. Leaf of tulsi and rhizome of *Zingiber officinale* Rosc. are crushed together. It is diluted in water. A spoonful of juice is administered for elders against cold and cough. It is advised for 2 to 3 days.
3. Leaves are chewed for cleaning mouth and throat.

Notes from Classics

Root decoction is given on malarial fever. Leaves have expectorant properties. Leaf juice is used on catarrah and bronchitis, also useful on skin diseases. Infusion of leaves is used as stomachic, in gastric disorders of children and hepatic affections. Leaf juice is carminative, febrifuge, refrigerants and laxative. Leaf powder is snuffed on ozaena. Leaf juice is dropped in ears for ear-ache. Seeds are mucilagenous and demulcent, are given in disorders of genito-urinirary problems. Flowers mixed with honey, ginger, and onion juice are given on cough and as expectorant. Juice of leaves, flower tops and slender roots of same plant is used as antidote on snake-bite. Plant is sacred to Hindu religion.

(Watt, 1889–1893)

Decoction of root is given as diaphoretic in malarial fever. Leaf juice possesses diaphoretic properties, used as stimulating, expectorant, also used in catarrah and bronchitis. It is used as an application to skin for ring-worms and cutaneous diseases. Juice is dropped in ears against ear-ache. Infusion of leaves is used as stomachic, used in gastric disorders of children. Seeds are given in disorders of genito-urinary system. Leaves are used as condiments in salad and other food, also used as pot-herb.

(Anonymous, 1948–1976)

Decoction of roots is diaphoretic, used in malarial fevers. Leaves are stimulant, diaphoretic, antiperiodic and expectorant, is used in catarrah, bronchitis, ring worms and cutaneous diseases. Infusion of leaves is used as stomachic. Leaves yield essential oil which has insecticidal and antibacterial properties, also said to be mosquito-repellant.

(Ambasta, 1986)

Leaves are considered as abortifacient, used in blood purification, in the treatment of cold, bronchitis and cough. Leaf juice is used on fever, ear complaints, gastric complaints, head-ache, intestinal germs, itch, leprosy, liver complaints, malaria, ring-worms and wounds. Seeds are used as cooling drink and considered as laxative. Entire plant is useful on cholera, constipation, diarrhoea, dropsy, protracted labour, vomiting and chicken-pox. Plant is considered as cooling drink. Plant is used as antidote on snake-bite. Wood is used for ornamental beads. Plant is hold sacred by Hindu religion.

(Jain, 1991)

**Opuntia dillenii* Haw.

Family: Cactaceae, Local Name: Phalisabar

Ethnobotanical Notes

Medicine

1. Stem of opuntia, stem bark of *Cordia dichotoma* Forst. *f.* and *Pterocarpus marsupium* Roxb. are crushed. About 20 ml of extract is given on menstrual complaints once a day till cure.

Notes from Classics

Poultice of leaves is said to allay heat and inflammation. Baked fruits are given in cough. Fruits are considered to be refrigerant. Syrup of fruits increases secretion of bile and controls spasmodic cough and expectoration. Leaf pulp is applied on eyes on opthalmia. Leaves are roasted and poultice is used on scorbutic ulcers. Plant juice is used as purgative and demulcent in gonorrhoea. Plant is useful as hedge or fence plant.

(Watt, 1889–1893)

Baked fruits are given in whooping cough. Fruits syrup is given to increase bile secretion and also controls spasmodic cough and expectorations. Internodes are made into pulp and applied on eyes for opthalmia.

(Anonymous, 1948–1976)

Mashed stem are used as poultice of allay inflammation and hot ones applied to boils to hasten suppuration and for poulticing guinea-worms and abscesses. Baked fruits are used for whooping cough. Fruit syrup increases secretion of bile and controls spasmodic cough and expectorations. Stem pulp is applied on opthalmia. Fruits are edible, also used in extraction of alcohol. Plant yields coarse fibre used for paper pulp. Plant is also used as hedge for farmyard.

(Ambasta, 1986)

Fruits are used on gonerrhoea and whooping cough. Fruits are edible. Plant is useful against pimples, sores and syphilis.

(Jain, 1991)

Critical Notes

Stem is not noted in past useful for menstrual complaints. Plant is exotic, native of South America.

**Oroxylum indicum* (L.) Vent.

Family: Bignoniaceae, Local Name: Tetu

Ethnobotanical Notes

Medicine

1. Stem bark powder is added in water. This infusion is given to cattle for 3–4 days to improve appetite.
2. Decoction of stem bark about 20 ml per day at night is given to regulate menstruation.
3. Stem bark of this plant and roots of *Abelmoschus tetraphyllus* (Roxb. ex Horn.) Borss. are crushed. This extract about 10–15 ml is given early in morning against leucorrhoea for a week. Consumption of food is avoided for 5 to 6 hours after administration of dose.

Notes from Classics

Roots have astringent and tonic properties and are employed in diarrhoea and dysentery. Decoction of bark is used in rheumatic swellings. Infusion of bark is most powerfully diaphoretic and slightly anodyne. Root paste mixed with turmeric powder is applied on sores on back of bullocks and horses.

(Watt, 1889–1893)

Young shoots and flowers are eaten as vegetable. Leaves are eaten by cattle. Thin light seeds are used as stuffing materials for hats and umbrellas. Bark and fruits are used as mordant in dyeing and tanning. Wood is used for match boxes and also used as fuel.

(Anonymous, 1948–1976)

Root bark is tonic and astringent, used in diarrhoea, dysentery and rheumatism, also considered as diaphoretic. Root bark boiled in oil is used in otorrhoea. Decoction of leaves is given in stomach-ache. Leaves are externally used for enlargement of spleen. Tender leaves are externally used for enlargement of spleen. Tender fruits are stomachic. Seeds are considered as purgative. Entire plant is said to be antiseptic. Leaves and shoots are eaten as vegetable. Leaves are used as cattle fodder. Wood is used for making match boxes.

(Ambasta, 1986)

Root bark is given on dysentery, diarrhoea, to induce conception, and for miscarriage. Roots are considered as tonic. Stem bark is useful for body and chest-pain, also useful on jaundice, oedema and scabies. Leaves are used against head-ache and stomach-ache. Seeds are purgative and used on piles. Plant is used in the treatment of colic, cholera, dyspepsia, epilepsy, fever, madness, pimples, small pox, sores and spleen complaints. Roots are used as antidote on snake-bite. Stem bark is applied on veterinary wounds. Flowers are eaten as vegetable. Fruits are eaten raw.

(Jain, 1991)

Critical Notes

Stem bark is effective to improve appetite, to regulate menstruation and to treat leucorrhoea. These are yet untapped.

**Ougeinia oogeinensis* (Roxb.) Hochr.

Family: Papilionaceae, Local Name: Tiwas

Ethnobotanical Notes

Medicine

1. Stem bark powder of same plant and also of *Ziziphus mauritiana* Lam. Non Mill. *Acacia nilotica* L. ssp. *indica* (Bth.) Brenan and fruit powder of *Terminalia chebula* L. are mixed in water overnight. Two spoons of this infusion is administered for the treatment of chronic cough.

Miscellaneous

1. Leaves are fed to domestic cattle especially in summer season.
2. Wood is used for breaking soil lumps in field, called 'Phali Ghalne'.

Notes from Classics

Bark is used in cases of diarrhoea and dysentery. Bark is said to be febrifuge and decoction of bark is used in dark coloration of urine. Leaves are lopped for cattle fodder. Wood is used for agricultural implements, carriage poles, wheels, furniture and building purposes. Bark is astringent, used as fish poison.

(Watt, 1889–1893)

Uses noted by Anonymous (1946–1976) and Ambasta (1986) are similar as noted by Watt (*loc.cit.*).

Stem bark is used to cure asthma, also controls dysentery and bronchitis in combination with leaves. Plant is useful in diarrhoea. Flowers are edible.

(Jain, 1991)

Critical Notes

Stem bark is used to treat chronic cough. It has not been tapped in past.

**Oxalis corniculata* Linn.

Family: Oxalidaceae, Local Name: Ambushi

Ethnobotanical Notes

Medicine

1. Laves are chewed to quench thirst.
2. Leaves are used as salad.

Notes from Classics

Leaves are considered cooling, refrigerant, stomachic and antiscrobutic. Fresh juice and leaves of same plant are reduced to paste with ghee and curdled milk is a useful medicine in diarrhoea, dysentery, prolapsus of rectum tympanitis, piles and difficult micturation. Curry of fresh leaves is given to improve the appetite and digestion of dyspeptic persons. Poultice of leaves is applied over inflamed parts. Leaves are useful in dysentery of children. Plant is cooked as pot-herb in times of scarcity.

(Watt, 1889–1893)

Leaves are considered cooling, refrigerant, stomachic, antiscrobtic and appetizing, used in fevers, dysentery scurvy and biliousness and for removing corns, warts, and other excrescences of skin. Infusion of leaves is used to remove opacities of the cornea. It is dropped into sore eye for itching lids. Plant possesses astringent, vermifuge, emmenogogue and antiseptic properties. Fresh juice is used to cure dyspepsia, piles, anaemia and tympanitis.

(Anonymous, 1948–1976)

Infusion of leaves is used to cure capacity of the cornea. Leaf juice is given to counteract to Datura Poison. Seeds are consumed in times of scarcity. Leaves are eaten as salad or cooked as vegetable or as pickles and chutneys.

(Ambasta, 1986)

Leaves are used as appetizer, refrigerant on cough, cuts, eye complaint, to improve digestion, to control dysentery and fever, used on rickets, stomach-ache, swellings, warts and eaten as salad. Entire plant is useful on epilepsy, jaundice, scurvy and wart. Plant is useful as antidote against snake-bite or poisonous insect-bite. Plant is cooked and eaten as vegetable.

(Jain, 1991)

Critical Notes

To quench thirst, the leaves are not reported yet. Plant is exotic, native of South Europe/North America.

Pancratium sanctae-mariae Blatt. & Hallb.

Family: Amaryllidaceae, Local Name: Pachan kand, Jungli kand

Ethnobotanical Notes

Medicine

1. Bulb juice, 2–3 drops, are poured in ears to relieve ear-ache.
2. Peels of bulbs are heated and kept on stomach to treat a round mass in stomach.

Miscellaneous

1. Stem of 'Kuda' (*Holarrhena pubescens* Wall ex Don) is placed in rice fields and bulb peels are placed on it. It is believed

that rice crop is prevented from diseases and keep away evil spirit. It helps for good produce. The practice is prevalent and followed by majority of tribal rice cultivators.

Critical Notes

The bulbs of this species are collected traditionally and employed especially by the tribal farmers. Now-a-days, they complain of their unavailability. It appears that these bulbs have been removed in such quantities that the plant is now in a precarious state. There is an urgent need to advise tribal farmers of this district. If proper measures are not taken, then this endemic species will become extinct in a very near future. Both applications are addition to the science of ethnobotany.

Pandanus odoratissimus Linn.*f.*

Family: Pandanaceae, Local Name: Kevada

Ethnobotanical Notes

Miscellaneous

1. Inflorescence bracts are buried in 'Nagali' (*Eleusine coracana* Gaertn.) fields to get better yield.
2. Male inflorescences are sold in local market being fragrant.

Notes from Classics

Oil obtained from bracts is considered stimulant and antispasmodic, administered on head-ache and rheumatism. Tender floral leaves are eaten raw or cooked with condiments. Drupe are edible.

(Watt, 1889–1893)

Male flowers yield Kevada attar and also used for decorating hair.

(Anonymous, 1948–1976)

Leaves are used for thatching huts, making mats, hats, baskets, fancy articles, umbrellas, etc. Leaf fibre is used for making sacks for coffee, sugar, grain etc. Leaves are used to treat leprosy, scabies, heart diseases, and brain complaints. Male flowers are used for extraction of Attar and Kevada water. Tender floral leaves are eaten raw or cooked. Anthers are used for the treatment of blood diseases.

(Ambasta, 1986)

Stem is used on ulcers. Roots are used as antidote on snake-bite. Leaves are used in syphilis, scabies, leprosy and skin diseases. Leaf fibres are used for fish nets. Leaves are used for thatching huts. Flowers are ornamentally used as hair-dye. Fruits are edible.

(Jain, 1991)

Critical Notes

Use of inflorescence of bracts in the crop of *E.coracana* for better yield is an ethnobelief.

Panicum miliaceum Linn.

Family: Graminae, Local Name: Warai

Ethnobotanical Notes

Medicine

1. Cooked grains are given to women after delivery for four to five days being better for digestion. It helps for good health.

Notes from Classics

Grains are used as nutritious food. Straw sometimes are used as cattle fodder.

(Watt, 1889–1893)

Grains are eaten boiled and cooked like rice grains. They are ground into flour for making chapattis. Grains are parched into 'Maraha' or 'Mard' used for making porridge and bread. Plant is excellent fodder for horses. Straw is used as bedding for cattle to protect from heat. Grains are fermented into liquor.

(Anonymous, 1948–1976)

Ambasta (1986) has recorded uses similar to those of Anonymous (*loc.cit.*).

Critical Notes

Consumption as food especially in post-delivery period keeps better health of a mother which in turn helps for better health of the child.

**Paracalyx scariosa* (Roxb.) Ali

Family: Papilionaceae, Local Name: Borvak

Ethnobotanical Notes

Medicine

1. Roots are kept overnight and infusion is given against stomach-ache. About two spoons is taken orally for 3–4 days.

Notes from Classics

Decoction of roots is given in dysentery. Plant is reported to be used as febrifuge, emmenogogue and analgesic in neuralgia.

(Watt, 1889–1893)

Plant is used as tonic, febrifuge, emmenogogue and analgesic. Decoction of roots is given on dysentery.

(Anonymous, 1948–1976)

Ambasta (1986) recorded similar uses as those in Anonymous (*loc.cit.*).

Roots are used against stomach ache and veneral diseases. Leaves are used as antidote on snake bite, also used as vermicide. Leaves are used in nasal blocks and cold.

(Jain, 1991)

Critical Notes

Roots are being reported for first time beneficial for stomach-ache.

**Pergularia daemia* (Forsk.) Chiov.

Family: Asclepiadaceae, Local Name: Kurmuda, Ran Karal

Ethnobotanical Notes

Miscellaneous

1. Leaves are used as fodder for goats.

Notes from Classics

Decoction of leaves is used in asthma. Leaf juice is given in infantile diarrhoea. Leaf juice is applied on rheumatic swellings. Leaf pulp is applied on carbuncles. Plant is emetic, expectorant and anthelmintic. Plant extract is given for uterine and menstrual troubles. It also facilitates parturition.

(Ambasta, 1986)

Leaves are used in the treatment asthma, catarrah, eye-diseases, diarrhoea, rheumatism, urine complaints, and also useful for wounds. Leaves in combination with roots is useful for head-ache. Flowers are used as a remedy on cough. Entire plant is used on asthma, blood purification, luecoderma and gastro-intestinal ulcers. Latex and fruits are used as veterinary medicine for calf-laying. Leaves are used as antidote on snake-bite.

(Jain, 1991)

Critical Notes

Fodder value is noted for the first time.

**Phaseolus lunatus* Linn.

Family: Papilionaceae, Local Name: Ran Udid

Ethnobotanical Notes

Medicine

1. Seed powder is given as anti-poison medicine which creates vomiting and reduces poison.

Notes from Classics

Young beans are cooked and eaten.

(Watt, 1889–1893)

Seeds are applied to abdomen of children in the form of poultice, and are astringent, used as diet in fever. Leaves are used in the form of pulp to colour pudding. Green vines are used as fodder and manure.

(Anonymous, 1948–1976)

Seeds are boiled or baked for table uses. Green vines are used as fodder, also made into hay. Sprouted seeds are boiled and eaten as vegetable. Crops grown as cover crop and green manure.

(Ambasta, 1986)

Critical Notes

The seeds are sold in markets by vendors. They help reduce poison. It is not recorded earlier.

**Phyllanthus fraternus* Webster

Family: Euphorbiaceae, Local Name: Dader

Ethnobotanical Notes

Medicine

1. Plant juice is applied on boils and on skin diseases.

Miscellaneous

1. Plants are used as cattle fodder.

Notes from Classics

Roots are used as excellent remedy for jaundice. Tender shoots are used as deobstruent, diuretic and healing. Tender shoots with fenugreek seeds are used on chronic dysentery. Entire plant is useful in dropsy, gonorrhoea, and other genito-urinary affections.

(Watt, 1889–1893)

Roots are useful in jaundice and also used as galactogogue. Decoction of leaves is used as refrigerant on scalp. Leaves and roots are made into poultice in rice water. It is used on oedema, swellings and ulcers. Poultice mixed with oil is used on opthalmia. Herb possesses astringent, deobstruent, stomachic, diuretic, febrifugal and antiseptic properties. It is used in stomach troubles, such as dyspepsia, colic, diarrhoea, dysentery and urinogenital complaints. Roots are fed to camels and as remedy for stomach-ache. Stem and leaves yield black cotton dye.

(Anonymous, 1948–1976)

Roots are given on jaundice and also used as galactogogue. Plant is considered as astringent, deobstruent, stomachic, diuretic, febrifuge, used in stomach complaints like diarrhoea, dysentery, dyspepsia dropsy and colic also used in urinogenital system. Latex is applied on sores. Leaves and stem yields cotton black dye.

(Ambasta, 1986)

Roots are abortifacient, used on indigestion and gastric diseases. Leaves are used on allergy, urino-genital complaints and gonorrhoea. Entire plant is useful on boils, diarrhoea, dysentery, dropsy, jaundice and to cure ulcers.

(Jain, 1991)

Critical Notes

Plant is useful for boils and skin diseases, also as fodder. These are unrecorded uses.

**Physalis minima* L.

Family: Solanaceae, Local Name: Phuga

Ethnobotanical Notes

Miscellaneous

1. Ripe fruits are edible.
2. Plant is used as fodder for cattle.

Notes from Classics

Fruits are considered as diuretic, purgative and tonic, used as a remedy in ulceration of bladder. Fruits are eaten by natives.

(Watt, 1889–1893)

Uses recorded by Anonymous (1948–76) and Ambasta (1986) are similar to those of Watt (*loc.cit.*).

Roots and leaves in combination are used on gastric diseases. Stem and leaves are used on stomach-ache. Entire plant is useful on ear-ache and abdominal pains. Fruits are edible.

(Jain, 1991)

Critical Notes

Its fodder value is still unnoted. Plant species is exotic, native of South America.

**Piliostigma foveolatum* (Dalz.) Thothathri

Family: Caesalpiniaceae, Local Name: Chamel

Ethnobotanical Notes

Medicine

1. Fresh bark is crushed and hands are washed by juice obtained as a remedy for jaundice treatment. It is done for 3 to 4 days.

Miscellaneous

1. Leaves are kept in containers of food grain to prevent from moisture and pests especially in monsoon season.

Critical Notes

Bark is important in jaundice. Leaves are useful to keep off the grains from paste. These utilities are still unnoticed.

**Piliostigma malabarium* (Roxb.) Benth.

Family: Caesalpiniaceae, Local Name: Shid

Ethnobotanical Notes

Medicine

1. Root paste is heated slightly, applied on dislocations to cure injury caused by agricultural tools.

Miscellaneous

1. Leaves are used by tribals for wrapping bidis.
2. Tender leaves are cooked and eaten as vegetable.

Notes from Classics

Tender leaves are eaten as a vegetable.

(Watt, 1889–1893)

Young shoots and leaves are eaten in Burma. Leaves are used as cattle fodder.

(Anonymous, 1948–1976)

Young shoots and leaves are eaten.

(Ambasta, 1986).

Bark is used on dysentery and loose-motions. Young shoots and flowers, are eaten as vegetable.

(Jain, 1991)

Critical Notes

Root paste to treat dislocations and injuries, and leaves as bidi wrappers are new applications.

**Pimpinella heyneana* Wall. ex Kurz

Family: Umbelliferae, Local Name: Dongarjira

Ethnobotanical Notes

Miscellaneous

1. Seeds are cooked and eaten as pulse (bhaji).

Notes from Classics

Roots are considered as a febrifuge.

(Ambasta, 1986)

Leaves and seeds are used as condiment.

(Jain, 1991)

Critical Notes

Edibility of seeds is unreported.

**Pistia stratiotes* Linn.

Family: Araceae, Local Name: Gondal

Ethnobotanical Notes

Medicine

1. 'Misri' *i.e.*, ash from leaves is mixed with coconut oil and applied on skin burns.

Miscellaneous

1. Leaf juice is given to goats suffering from throat swellings.

Notes from Classics

Leaf are boiled in coconut oil and applied externally in chronic skin diseases. Leaves are cooked and eaten, valued as manure and applied to ring-worms of the scalp.

(Ambasta, 1986)

Critical Notes

Leaf juice to treat throat swellings and leaf ash for skin burns are not yet reported.

Pithecelobium dulce (Roxb.) Bth.

Family: Mimosaceae, Local Name: Vilayati-chinch

Ethnobotanical Notes

Miscellaneous

1. Seed arils are eaten by natives, cherished especially by children.
2. Leaves are lopped for fodder for domestic cattle.
3. Green fruits are also fed to goats.

Notes from Classics

Pulpy arils are eaten by poorer classes. Wood is used for making carts, packing boxes, and panels of doors.

(Watt, 1889–1893)

Stem bark is astringent, used in dysentery and as febrifuge. Leaves are lopped for goats, sheep, horses, and cattle as fodder. Wood is used for general constructions, packing cases, fence posts, carts, and agricultural implements, also used as excellent fuel.

(Anonymous, 1948–1976)

Seeds are eaten raw or in curries. Leaves and pods are eaten as fodder. Wood is used for general construction.

(Ambasta, 1986)

Roots are antirabic, used against dog-bite. Leaves are used in dropsy also promote hair growth. Fruits are edible.

(Jain, 1991)

Critical Notes

Uses noted by investigator are noted through out the country. Plant species is exotic, native of Mexico.

******Phoenix sylvestris*** (L.) Roxb.**

Family: Arecaceae, Local Name: Shindi, Tadi

Ethnobotanical Notes

Medicine

1. Stem trunk is given a cut at apex and the sap called 'Nira' is collected in earthern pots early in the morning before sunrise. It is drunk as a nutritious drink. It is also thought useful for kidney complaints and urinal obstructions.

Miscellaneous

1. Ripe fruits are eaten.
2. Leaves are used for making brooms to sweep houses, huts and fore-grounds.
3. Dried trunk is used as fire wood.
4. Trunk is also used as pole for cattle shed.

Notes from Classics

Roots are used on tooth-ache. Sap obtained from tree is considered as cooling beverage. Paste of kernels and roots of *Achyranthes aspera* L. are chewed with betel leaves as a remedy for ague. Central tender part is used on gonorrhoea and gleet. Fruits are eaten by natives. Leaves are used for making mats, baskets, bags, brooms, fans, also used for paper making. Petioles are beaten and twisted into ropes.

(Watt, 1889–1893)

Sap after it has undergone spontaneous fermentation for 8–10 hrs. forms a common beverage (Tadi). It is also used to limited extent for making vinegar. Palm molasses is used for edible purposes and is a source of alcohol. Fruits are edible, preserved into jams and jellies. Leaves are used for thatching huts and for making mats, hats, fans, baskets, bags, fishing nets etc. Petioles are beaten into ropes. Female spadix forms a good brush for white wash. Leaves are lopped for fodder. Wood is used for temporary construction, bridges, tents, and pegs.

(Anonymous, 1948–1976)

Leaves are lopped for fodder, also used for thatching, making mats, baskets, brooms, fishing nets etc. They also yields soft fibre. Fruits are edible, used in jams and jellies. Sap tapped from tree is used for making jaggery and sugar. Sap called 'Nira' is a refreshing drink. After fermentation it form 'Toddy'. Trunk freed from pith is used as water conduits.

(Ambasta, 1986)

Roots are used in gum trouble and tooth-ache, menstrual complaint and veneral diseases. Stem is used in gonorrhoea. Leaves are used for basketery, used in diarrhoea and dysentery. Fruits are used in beverages.

(Jain, 1991)

Critical Notes

Uses noted by the present investigators are also widely documented throughout the country. It can be recommended for captive plantation especially along river banks, rivulets or water reservoirs, being important from viewpoint of cottage industry.

**Plectranthes mollis* (Ait.) Spr.

Family: Labiatae, Local Name: Jungli-tili

Ethnobotanical Notes

Miscellaneous

1. Leaves are crushed or rubbed on stone, it is mixed in pond water for fish killing.
2. Fresh leaves are kept in clothes and blankets to prevent from fleas, ticks, termites etc.

Notes from Classics

Leaf juice is used to stop bleeding, and also used to treat fever. Leaves are used as mosquito-repellant.

(Anonymous, 1948–1976)

Uses recorded by Ambasta (1986) are similar as recorded by Anonymous (*loc.cit.*).

Seeds are used as tonic, also used on rheumatism.

(Jain, 1991)

Critical Notes

Leaves are being reported for the first time to stupefy fishes. Its applications as repellent for fleas, ticks and termites are new.

Plumbago zeylanica L.

Family: Plumbaginaceae, Local Name: Chitrak

Ethnobotanical Notes

Medicine

1. Roots are rapped in cloth. The wheel of hand-mill is kept on it adjusting the central whole of the wheel. It is then ignited from below. The person suffering from piles is advised to sit on the wheel in such a way that the place of piles and the central hole of the wheel are adjusted. The smoke comes in contact with the piles and helps to cure it.

Notes from Classics

Roots are useful in the treatment of dyspepsia, piles, anasarca, diarrhoea, skin diseases, promotes appetite and also increases digestive powers. Root paste is externally applied on skin diseases and cures leprosy.

(Watt, 1889–1893)

Roots are abortificient and vesicant, also considered as diuretic, caustic and expellant, useful to treat rheumatism, as irritant to skin, in the treatment of dyspepsia, piles, anasaraca and diarrhoea. Root paste is applied on open abscesses. Paste mixed with milk, vinegar or salt water is externally applied on leprosy and skin diseases, also apt to cause abortion. Infusion of roots is used on influenza and black water fever in Africa. Tincture of root bark is said to be antiperiodic.

(Anonymous, 1948–1976)

Uses recorded by Ambasta (1986) are similar as given in Anonymous (*loc.cit.*).

Roots are used against abdominal pain, used as abortifacient, in antifertility, aphthae, appetite, controls diarrhoea, dysentery, dyspepsia, fever, eczema, head-ache, leprosy, improves indigestion, in the treatment of leucoderma, muscle pain, paralysis, piles, rheumatism and skin diseases. Roots are used as veterinary medicine for skin applications. Roots are symbolically worn on arms to get relief from body pain. Twigs and leaves are used in malaria. Plant is useful in spleen complaints.

(Jain, 1991)

Critical Notes

Plant species is exotic, native of Africa.

**Plumeria rubra* L.forma *acuminata* (Ait.) Santapau & Irani ex Shah

Family: Apocynaceae, Local Name: Pandhra-chapha

Ethnobotanical Notes

Medicine

1. Stem bark is heated and kept on abdomen for 8–10 days. It is believed that it causes impotency, which does not require operation of family planning.

Miscellaneous

1. Infusion of bark if sprayed on a honey comb, honey bees run away without harming the honey collector.

Notes from Classics

Bark is purgative and used in cases of leprosy. Bark is recommended as a cure for gonorrhoea. Leaves are made into poultice, used for swellings. Milky juice is employed as a rubefacient in rheumatism. Blunt-ended branches are introduced into uterus to procure abortion. Flower buds are eaten with betel leaves in ague.

(Watt, 1889–1893)

Root is violent cathartic. Decoction of bark is used as purgative, emmenogogue and febrifuge. Latex of plant has rubefacient and purgative properties. It has entered into applications for itch, rheumatism and gum troubles. Wood is used for making drums and other musical instruments.

(Anonymous, 1948–1976)

Ambasta (1986) have recorded similar uses those recorded by Anonymous (*loc.cit.*).

Roots are abortifacient. Stem bark is used in malarial fever. Leaf is used to increase lactation. Plant is used in the treatment of cholera and improves idigestion.

(Jain, 1991)

Critical Notes

Stem bark is important medicine to reduce sex in both men and women. Its infusion is used while collecting honey. Both application are new to the science of ethnobotany.

Pogostemon benghalensis (Burm.*f.*) O.Ktze.

Family: Labiatae, Local Name: Phangli, Fangla

Ethnobotanical Notes

Medicine

1. Leaf paste is applied on wounds and injuries. It is used as antiseptic.

Notes from Classics

Ash of stem is used as manure for paddy crop.

(Ambasta, 1986)

Leaf paste is applied on wounds and cuts. Entire plant is used on hemorrhagia. Whole plant is used veterinary medicine on wounds.

(Jain, 1991)

***Pogostemon parviflora* Bth.**

Family: Labiatae, Local Name: Phangula

Ethnobotanical Notes

Medicine

1. Half cup of decoction of roots is useful for adults against stomach-ache. It is taken orally only once.
2. Roots and leaf juice are given as antidote against snake bite.
3. Fresh leaves are kept below cap on head to prevent from sun-stroke.
4. Leaves of this plant and *Vitex negundo* L. are crushed and extract is kept between jaws to get relief from tooth-ache.
5. Twigs are boiled in water. This water is used for bath to get relief from stomach-ache.

Notes from Classics

Leaves are used to wash wounds. Roots are remedy against bite of 'Phursa Snake" (*Echis carinata*) roots are chewed. Leaves were eaten in Khandesh during famine of 1877–1878.

(Watt, 1889–1893)

Roots are stimulant and antihaemorrhagic. Bruised leaves are applied as cataplasm to wounds. Leaves are used as green manure. Leaves are eaten in times of scarcity, used as a stimulant and styptic.

(Anonymous, 1948–1976)

Ambasta (1986) recorded similar uses as those recorded by Anonymous (*loc.cit.*).

Roots are used on abdominal pain after child-birth.

(Jain, 1991)

Critical Notes

Roots are beneficial for stomach-ache. Leaves are useful in treating sun-stroke and tooth-ache. All these applications are additional and not documented yet.

Polygonum barbatum Linn.

Family: Polygonaceae

Ethnobotanical Notes

Medicine

1. Extract of leaves are applied on septics and wounds.

Notes from Classics

Roots are astringent and cooling. Decoction of shoots is used as stimulating wash for ulcers, stem juice acts as cicatrizent. Powdered leaves are applied to fly-infected wounds of goats. Seeds posses tonic, purgative and emetic properties, also used as colic.

(Anonymous, 1948–1976)

Decoction of shoots is used as stimulating wash for ulcers, juice acts as cicatrizent. Seeds are tonic, purgative, and emetic.

(Ambasta, 1986)

Critical Notes

Leaves are found useful for the first time for septics and wounds.

Polygonum plebeium R.Br.

Family: Polygonaceae, Local Name: Bose

Ethnobotanical Notes

Medicine

1. One spoon of leaf juice is given to children for 3 to 4 days as vermicide.

Notes from Classics

Root is given in bowel complaint.

(Watt, 1889–1893)

Plant is used as vegetable. Plant is liked by horses. Plant is used as vegetable. Plant is useful in bowel complaints and pneumonia.

(Ambasta, 1986)

Root and leaves are used on cough and indigestion. Entire plant is useful on diarrhoea, dysentery, gastric complaints, lactation and as vegetable.

(Jain, 1991)

Critical Notes

Medicinal utility as vermicide is not reported earlier.

******Pongamia pinnata*** (L.) Pierre.**

Family: Papilionaceae, Local Name: Karanj

Ethnobotanical Notes

Medicine

1. Seed oil is used as an application for eczema and ring worm.
2. Leaves and bark are heated and the fumes are passed over the swellings or inflammations.

Notes from Classics

Root juice mixed with coconut milk and lime water is remedy for gonorrhoea. Root and stem oils have antiseptic properties. Leaf poultice is applied for ulcers infected with worms. Leaves are given internally in leprosy. Leaf juice is used on flatulence, dyspepsia and diarrhoea. Tender leaves are recommended for piles. Flowers are used in the treatment of diabetes. Pods are worn round the neck to prevent from whooping cough. Seed oil is used on skin diseases and rheumatism. Seed pulp is useful on leprosy. Leaves are used as cattle fodder. Wood is used as fuel for oil mills.

(Watt, 1889–1893)

Root juice is used for cleaning foul ulcers, closing fistulous, cleaning teeth and strengthening gums, also used in gonorrhoea. Root paste is applied on scrofulous enlargement. Twigs are used for brushing teeth. Stem bark is given internally on bleeding piles. Decoction of bark is given on beri-beri. Leaf juice is prescribed in flatulence, dyspepsia, diarrhoea and cough, also useful in leprosy and gonorrhoea. Hot infusion of leaves is medicated bath for relieving rheumatic pains. Decoction of flowers is used to quench thirst in diabetes. Leaves are lopped for fodder, acts as galactogogue.

Seed oil is used as lubrication, illumination and other medicinal purposes. Seed cake is fed to livestock, also used as manure for coffee and tomato. Roots are used as fish poison. Wood is used for making bullock-carts, ploughs, furniture, turnery articles, paper pulp, plywood etc., also used for thatching huts. Wood ash is used as dye.

(Anonymous, 1948–1976)

Uses noted by Ambasta (1986) are similar as recorded in Anonymous (*loc.cit.*).

Roots are used on fistula, sores and ulcers. Stem bark is used in malaria and to treat piles, skin diseases and wounds. Twigs are used as tooth brush. Leaves are applied on skin diseases. Flowers are used in diabetes. Fruits are used to treat cough. Seeds are used on bronchitis, burns, cold, cough, ear-complaint, eczema, fever, itching, joint pain, rheumatism, scabies and whooping cough. Seeds are edible. Seed oil is used as fish poison and also used as bird-lime. Plant is useful in application for leprosy.

(Jain, 1991)

Critical Notes

Seed oil applied for eczema and ringworm is additional. Leaves and the plant itself are being noted for the first time to treat swelling and inflammation.

Portulaca oleracea L.

Family: Portulacaceae, Local Name: Ghatghol.

Ethnobotanical Notes

Miscellaneous

1. Leaves are mixed with pulse Tur [*Cajanus cajan* (Linn.) Millsp.] or Udid [*Vigna radiata* (Linn.) Wilczek.] dal and cooked as vegetable.

Notes from Classics

Bruised fresh leaves are used as external application for erysipelas and infusion of it also said to be diuretic in dysuria, used in burning fevers. Leaves are used as remedy for head-ache. Seeds are recommended in kidney complaint, bladder and lung diseases caused by hot or bilious humours. Seed powder is externally pressed on burns and scalds and various skin diseases. Seeds are useful on mucous membrane intestinal canal, dysentery and mucous diarrhoea. Entire plant is useful in blood purification, applied on burns, against fevers, prickly heat and burning sensation. Leaves are used as vegetable.

(Watt, 1889–1893)

Plant is considered as refrigerant, vulnerary, anti-scrobutic aperient, and diuretic. Useful in scurvy, liver diseases, spleen, kidney and bladder complaints. It is also employed in cardio-vascular diseases, dysuria, haematuria, dysentery, sore nipples and ulcerations of mouth. Roasted seeds are eaten, considered as diuretic, antidysentric, and externally applied on burns and scalds. Plant is used as pot-herb, consumed as salad and employment in soups, fleshy stems are pickled. Plant is used as fodder for cattle, sheep and pigs. Plant is used as green manure.

(Anonymous, 1948–1976)

Ambasta (1986) recorded similar uses as those recorded by Anonymous (*loc.cit.*).

Leaves are used in the treatment of head-ache. Plant is used for blood for purification, applied on burns, also used against fever, prickly heat and burning sensation.

(Jain, 1991)

Critical Notes

Plant species is exotic, native of North Africa.

Portulaca quadrifida Linn.

Family: Portulaceae, Local Name: Ghol

Ethnobotanical Notes

Miscellaneous

1. Plants are used as vegetable.
2. Plants are also used as cattle fodder.

Notes from Classics

Infusion of leaves is diuretic, refrigerant, astringent, emollient and used in dysuria. Fresh leaves act as ice-cold lotion. Seeds and leaves are useful in haemoptysis and gonorrhoea. Seeds are given in cases of dysentery and mucous diarrhoea. Plant is used as pot-herb.

(Watt, 1889–1893)

Leaf paste is an application on scalds, burnings and erysipelas. Herb is considered to possess refrigerant, vulnerary, anti-scorbutic, aperient and diuretic, properties, useful in the treatment of scurvy, liver, spleen, kidney and bladder complaints, in the treatment of cardio-vascular diseases, dysuria, hematuria, gonorrhoea, dysentery, sore nipples and ulcers of mouth. Plant juice is used on tooth-ache and ear-ache. Herb is used as vegetable. Roasted seeds are diuretic and antidysentric. Seed paste is used on burns and scalds.

(Anonymous, 1948–1976)

Herb is used in asthma, cough, urinary discharge, inflammations and ulcers. Herb is eaten as vegetable.

(Ambasta, 1986)

Leaves are used as antiscorbutic, used on burns, ulcers and as cooling agent. Plant is used on skin diseases. Plant is eaten as vegetable.

(Jain, 1991)

Prosopis cineraria (L.) Druce

Family: Mimosaceae, Local Name: Saundhal

Ethnobotanical Notes

Miscellaneous

1. Leaves and pods are used as cattle fodder.

Notes from Classics

Bark is locally employed as a remedy for rheumatism. Fruits are consumed by poorer classes, cooked green or dry. Pods are good fodder for goats, sheep, cattle and camels. Wood is used for building carts, well curbs, furniture and agricultural implements, also valued as fuel.

(Watt, 1889–1893)

Bark is ground and flour is made into cattle which are consumed by natives. Bark and leaf galls are used for tanning. Leaves and pods are lopped for fodder and as green manure. Flowers mixed with sugar are administered to prevent miscarriage. Pods are eaten green in times of scarcity. After boiling, they possess astringent, demulcent, and pectorial properties. Wood is suitable for interior construction work cart wheels and agricultural implements, also used as fuel and for charcoal making.

(Anonymous, 1948–1976)

Bark is used as food in famine periods. Pods are eaten raw, green or dry. After boiling, they are astringent, demulcent and pectoral. Wood is suitable for construction, wheels, hubs of carts, agricultural implements, tool handles, etc. Wood is used as fuel and charcoal making.

(Ambasta, 1986)

Roots and flowers are used to treat miscarriage. Bark, gum and shoots are edible. Fruits are eaten as vegetable. Sikh Saint Gurunanak planted many trees hence thought sacred. Pandavas concealed arms under this tree.

(Jain, 1991)

**Prosopis juliflora* DC.

Family: Mimosaceae, Local Name: Vilayati Babhul

Ethnobotanical Notes

Miscellaneous

1. Leaves are used as goat and sheep fodder.
2. It is used as fuel.
3. Dried branches are used for fencing farmyards.

Notes from Classics

Leaves are used as fodder for live-stock. Ripe pods are used as stock feed and staple food after removal of seeds. Seed powder is used for preparing bread. Gum is used as emulsifying agent, also used in confectionary. Wood is used for house building, furniture, turnery, as fencing posts, hard board sheets etc.

(Ambasta, 1986)

Fruits are eaten to quench thirst.

(Jain, 1991)

Critical Notes

Its fuel value is not noted in past. It is exotic, naturalized fast on waste lands. Although troublesome, it is useful as a source of fuel especially for poorer classes. It is fast growing.

Psoralea corylifolia L.

Family: Papilionaceae, Local Name: Bauchi

Ethnobotanical Notes

Medicine

1. Seed oil is applied on leucoderma.

Miscellaneous

1. Entire plant is used as fodder plant for cattle.

Notes from Classics

Roots are used in cures of teeth. Leaves are used in the treatment of diarrhoea. Seeds are laxative, aphrodisiac, anthelmintic, diuretic and diaphoretic infebrile condition. It is especially recommended in the treatment of leucoderma, lepsosy, psoriasis, and inflammatory diseases of skin. Seed powder is orally administered on head-ache. Plant is used as fodder and green manure.

(Watt, 1889–1893)

Uses recorded by Ambasta (1986) are similar to those of Anonymous (*loc.cit.*).

Seeds paste is used on itching, leprosy, leucoderma, psoriasis, skin diseases and injury. Plant is used as veterinary medicine.

(Jain, 1991)

**Pterocarpus marsupium* Roxb.

Family: Papilionaceae, Local Name: Bivala, Lalbivala

Ethnobotanical Notes

Medicine

1. Stem bark is boiled and decoction about 20 ml is given to drunk for a week to control excessive menstruation.

2. Infusion of stem bark is given to teenager girls against menstrual complaints.
3. Red latex is kept on tongue of a person suffering from bleeding stools.
4. Red gum is mixed with rice grains and given to anaemic persons.

Veterinary Medicine

1. Leaf extract is applied on skin infections of cattle.

Notes from Classics

Bruised leaves are used externally on boils, sores and skin diseases. Leaves are used as fodder for cattle and goats. Wood is used for doors, windows, furniture, agricultural implements, boats, railway sleepers etc.

(Watt, 1889–1893)

Stem bark is astringent, used on tooth-ache. Bruised leaves are extensively applied on sores and skin diseases. Infusion of wood is used in diabetes. Gum is powerfully astringent, used in diarrhoea and dysentery, externally applied on leucorrhoea and hemmorages. Timber is used for building purposes, railway wagons, carts, boats, agricultural implmenents, drums, tool handles, furniture, carving, wrapping paper pulp and also used as fuel.

(Anonymous, 1948–1976)

Uses recorded by Ambasta (1986) are similar to those of Anonymous (*loc.cit.*).

Stem bark is used on dysentery, as post-natal tonic, to cure tooth-ache and urine complaints. Leaves are used on boils sores and skin diseases. Wood is useful in asthma, body pain, chest pain and diabetes. It improves digestion. Drums are prepared from wood. Entire plant is used as a remedy to burns, cholera, menorrhoea syphilis, stomach-ache etc. Plant is used as veterinary medicine.

(Jain, 1991)

Critical Notes

Application of stem bark to treat menstrual complaints, of latex for bleeding stools and of gum for anaemic condition are additional.

*Pueraria tuberosa (Willd.) DC.

Family: Papilionaceae, Local Name: Bendar kand, Kohalvel

Ethnobotanical Notes

Medicine

1. Flour obtained from dried tubers is added in the flour of bread as a general body tonic.
2. Tubers are eaten as cooling.

Veterinary Medicine

1. Tubers mixed in fodder are fed to animals for flatulence.

Miscellaneous

1. Stem is splitted and fibre called 'Wak' is used to tie fodder and fuel wood bundles.

Notes from Classics

Tuber poultice is applied to reduce inflammatory swellings of joints. Tubes are given as tonic with milk.

(Watt, 1889–1893)

Roots are demulcent and refrigerant in fevers, used as cataplasm for swelling of joints and lactogogoue. Leaves are used as fodder for horses and cattle.

(Anonymous, 1948–1976)

Uses noted by Ambasta (1986) are similar as reported in Anonymous (*loc.cit.*).

Roots are used on body-ache and fever. Tubers are used as antiemetic, used in asthma, cholera, to treat chest-pain, diarrhoea, fever, rheumatism, swelling, also used as tonic, given to increase lactation, also used on abdominal pain. Seeds are useful on ulcers. Plant is used on mennorrhoea and syphilis.

(Jain, 1991)

Critical Notes

Edibility and tonic property of tubers is widely documented. Use of stem as fibres is new. Removal of tubers appear to have greatly affected the populations of this taxon.

**Pupalia lappacea* (L.) A.L.Juss.

Family: Amaranthaceae

Ethnobotanical Notes

Miscellaneous

1. Plant is used as cattle and goat fodder.

Notes from Classics

Fruit paste is applied on cuts. Fruit is ingredient of enema preparations, mixed with palm oil is applied as dressing for boils. Fruit soup is given for cough and fever. Ashes of burnt plant is taken with water for flatulence. Ash is applied on bleeding sores is leprosy. Plant is employed as bait to catch fishes.

(Anonymous, 1948–1976)

Fruit is ingredient of enema preparations, mixed with palm oil is applied to boils. Fruit soup is given for cough and fever.

(Ambasta, 1986)

Critical Notes

Utility of the plants for goat fodder is yet to be communicated.

**Radermachera xylocarpa* (Roxb.) K. Schum.

Family: Bignoniaceae, Local Name: Kharsing

Ethnobotanical Notes

Miscellaneous

1. Tender fruits are cooked and eaten as vegetable.

Medicine

1. Decotion of fruits and stem are administered against cold, fever and bodyache, headache.

Notes from Classics

Resin extracted from wood is used on skin troubles. Tender fruits are eaten as vegetable. Wood is used for house constructions, furniture, carts, carriages, agricultural implements, paneling and packing boxes.

(Ambasta, 1986)

Critical Notes

Applications of fruits and stem against cold fever, bodyache, headache additional.

Rhus sinuata Thunb.

Family: Anacardiaceae, Local Name: Amoni

Ethnobotanical Notes

Miscellaneous

1. Unripe fruits are cherished by children.
2. Leaves are used as goat and cattle fodder.

Notes from Classics

Bark is used for tanning. Wood is employed for fuel.

(Watt, 1889–1893)

Leaves are cooked and eaten. Fruits are cooked or roasted and eaten. Leaves are used as cattle fodder. Wood is used for fence post and fuel.

(Anonymous, 1948–1976)

Stem bark is used for tanning. Leaves are used for fodder. Fruits are eaten cooked or roasted. Wood is used for small articles and fencing.

(Ambasta, 1986)

Fruits are edible.

(Jain, 1991)

Rhyncosia minima DC.

Family: Papilionaceae

Ethnobotanical Notes

Miscellaneous

1. Plant is used as cattle and goat fodder.

Notes from Classics

Plant is reported to be eaten by cattle.

(Watt, 1889–1893)

Leaves are reported as abortifacient. Leaves are used as cattle and horse fodder.

(Anonymous, 1948–1976)

Leaves are used as an abortifacient. Plants are used as fodder for goats, cattle and horses.

(Ambasta, 1986)

Leaves are abortifacient, used on asthma and piles.

(Jain, 1991)

**Ricinus communis* Linn.

Family: Euphorbiaceae, Local Name: Erand

Ethnobotanical Notes

Medicine

1. Decoction of root is given twice a day against yellow and burning urination. One spoonful is taken orally till cure.

Notes from Classics

Stem bark is used for paper making. Seed oil is used for lamps, for lubricating machinery, clocks, watches, also used in railways. Oil is used for soap, candles, pomatums, dyeing clothes and leathers.

(Watt, 1889–1893)

Decoction of root is administered for lumbago and allied complaints. Root paste is applied for tooth-ache. Poultices of leaves are applied on sores, boils and swellings. Infusion of leaves is used against stomach-ache and as a lotion for eyes. Leaf powder gives relief on caries, and also applied for guinea-worms and sores. Leaves are considered as lactogogue and applied in the form of poultice over breast. Leaf juice is said to emetic, also useful in jaundice. Castor cake is used as manure. Dried branches are used for thatching hut walls. Dried seeds and seed hulls are used as highly combustible fuel. Castor oil is used in paint, enamels, varnishes, fabrics, flypaper, printing greases, polish, waxes.

(Anonymous, 1948–1976)

Poultice of leaves is applied on sores and boils. Decoction of roots is used on plumbago. Dried stems are used for thatching and as wattle in walls of mud huts, also used as fuel. Leaves are fed to cattle as fodder. Seed oil is used for lubrication and illumination, used for making soaps, fly paper, imitation leather, also used in polishes, candles, varnishes, plastic moldings, insulation adhesives. Oil is used for preservation of fruits and vegetable. Cake is used as manure.

(Ambasta, 1986)

Roots are used on lumbago and rheumatism. Leaves are used on boils, tooth-ache, head-ache, gum trouble, hydrocele, to cure injuries, joint pains, night blindness, pneumonia, rheumatism, skin diseases, swellings and wounds. Fruits are useful against sun-stroke, fruit paste is applied on foots. Seeds are given in constipation, contraception, against intestinal worms, seeds are laxative, treated in pneumonia. Plant is used on anasarca and burns, against cholera, delirium, dropsy, muscle pain, and to cure sores. Seed oil is used as vermicide, and also useful on scorpion bite.

(Jain, 1991)

Critical Notes

Roots are not reported earlier to treat yellow and burning urination. Plant is exotic, native of Africa.

Rivea hypocrateriformis (Desv.) Choisy.

Family: Convolvulaceae, Local Name: Phangvel

Ethnobotanical Notes

Miscellaneous

1. Leaves are cooked and eaten as vegetable.
2. Leaves are fed to cattle as fodder.

Notes from Classics

Leaves are used as fodder for cattle.

(Watt, 1889–1893)

Root is given to women after child-birth in Bihar. Leaves and young shoots are eaten as vegetable.

(Anonymous, 1948–1976)

Leaves are used as abortifacient. Plant is used as fodder foı cattle and horses.

(Ambasta, 1986)

Leaves are said to be abortifacient, used on asthma and piles.

(Jain, 1991)

**Rungia repens* L.

Family: Acanthaceae

Ethnobotanical Notes

Medicine

1. Leaves of this plant and *Mallotus philipiensis* (Lam.) Muell-Arg. and *Piper longum* Linn. are crushed together. Half cup of juice is given for 4–5 days to regularize unperiodic menstruation.

Notes from Classics

Leaves are bruised fresh, mixed with castor oil and applied on scalp in cases of *Tinea capitis*. Whole plant is dried and pulverized and given in fevers and cough, also considered and vermifuge.

(Watt, 1889–1893)

Herb is dried, pulverized and used in treatment of cough and fever. Fresh bruised leaves are mixed with castor oil and applied on scalp to cure *Tinea capitis*, a scurry fungoid infection occurring in children.

(Anonymous, 1948–1976)

Fresh bruised leaves are mixed with castor oil applied to scalp as a cure to *Tinea capitis* a scurry fungoid infection.

(Ambasta, 1986)

Critical Notes

To regularize menstruation, leaves are being reported useful for the first time.

Salmalia malabaricum (DC.) Schott & Endl.

Family: Bombacaceae, Local Name: Sawar

Ethnobotanical Notes

Medicine

1. Root paste is applied on injury caused by agricultural tools. It acts as an antiseptic.

2. Bark infusion, about half cup, is given orally in case of diarrhoea or bark is simply chewed and the extract is swallowed to control dysentery in children.

Notes from Classics

Decoction of root is used as tonic and stimulant. Tap root of young plant is used in gonorrhoea. Bark and roots are emetic. Inner bark of tree is suitable for cordage. Leaf paste is externally used for glandular swellings. Flowers are used as demulcent. Flowers alongwith poppy seeds, goats milk with sugar are boiled and given three times a day in haemorrhoids. Leaves and twigs are lopped for fodder. Seed yield soft and silky fibre used for stuffing pillows.

(Watt, 1889–1893)

Seed oil is used for soup making and as illuminant. Oil cake is fed to cattle. Young cake is fed to cattle. Young roots, tender leaves, flower buds, fleshy calyces and gum is eaten. Younger twigs are used as fodder. Floss of seeds is called as kapok, used for making life belts, stuffing cushions and pillows, insulating material for refrigerator, sound proof covers etc. Timber is used for match industry, and in manufacturing of cheap plywood, ceilings and toys etc.

(Anonymous, 1948–1976)

Ambasta (1986) recorded similar uses as those given by Anonymous (*loc.cit.*).

Root bark is useful in abdominal pain, cough and impotency. It is also applicable on burns. Bark paste is applied on acne pimples, boils and bone fracture. Bark is used against blood dysentery, to improve sex weakness, skin diseases, to treat urine complaints, vitality and also used as a tonic. Stem is used in syphilis, night pollution and to treat gonorrhoea. Leaves are used as edible vegetable, to treat anaemia and antifertility. Gum is used in delirium, diarrhoea, dysentery and asthma. Bark paste is applied on wounds and skin infections. Gum is used to treat acute dysentery, haemoptysis, of pulmonary tuber-culosis, influenza and mennorhagia. Gum is mixed with ashes and castor oil which is used as a cement for caulking iron pans for boiling sugar. Seeds are used in the treatment of chicken-pox and small-pox. Plant is useful in carbuncle, to treat cholera, gum trouble, hematuria, leprosy, scabies, sores in mouth, spermatorrhoea, spleen complaints, tooth-ache, and also applied on ulcers. Plant is used as antidote on scorpion-bite and snake-bite.

(Jain, 1991)

Santalum album L.

Family: Santalaceae, Local Name: Chandan

Ethnobotanical Notes

Miscellaneous

1. Wood pieces are burnt in 'Yagna' for happy and prosperous life.
2. It is believed that if wood pieces are burnt in pyre, the soul of a dead person reaches the heaven.

Notes from Classics

Emulsion of wood is used as cooling application to skin erysipelas prurigo and sudamina. Wood oil is used internally as a remedy for gonorrhoea. Wood rubbed up with water is used as an application in prickly heat. Wood paste is used externally to local inflammations on temples, in fevers, and to skin diseases, to allay heat and pruritus. It also acts as diaphoretic. Wood is described as cold and dry, cardiac, tonic, astringent, anti aphrodisiac and aleximorphic. Sandal wood is used for carving various articles and handicrafts.

(Watt, 1889–1893)

Wood oil is used to treat gonorrhoea and also to treat skin diseases. Wood paste is applied on burns, in fevers, and head-ache, applied on forehead and eyelids. Sandal wood is used as fixative in soap industry and toilet powders, cosmetics and making attars.

(Anonymous, 1948–1976)

Oil is described as diuretic, diaphoretic, refrigerant and expectorant. Wood oil is used in perfumeries, aggarbattis and attars. Wood is used for finest carvings.

(Ambasta, 1986)

Seeds are abortifacient. Wood is used in fevers, head-ache, skin diseases. Plant is used in dysentery and spleen complaints. Tree is symbolically sacred.

(Jain, 1991)

Critical Notes

The wood finds place in the ethnobeliefs in this district of Maharashtra.

Sapindus laurifolius Vahl

Family: Sapindaceae, Local Name: Ritha

Ethnobotanical Notes

Miscellaneous

1. Fruits are used for hair washing and also prevent hair fall.
2. Fruits are used for washing clothes.

Notes from Classics

Root is said to be expectorant. Three of four grains may be given by nose, in all kinds of fits, producing insensibility. Fumigations with it are useful in hysteria, melancholy and epilepsy. Seeds are used to stimulate uterus in child-birth and amenorrhoea. Fruits pulp is squenzed in cows milk and dropped in nostrils as the treatment in coma and dropped into eyes in delirium. One miskal (kernel) rubbed in water until it soap's and then strained, is given as antidote for snake bite and scorpion bite. Wood is used for house-building and construction of carts, and variety of small articles.

(Watt, 1889–1893)

Fruits are expectorant and have emetic properties, used in treatment of excessive salivation, epilepsy and chlorosis. Seed powder is used in treatment of dental caries. Leaves are lopped for fodder. Soapnuts are used as detergents for washing cloth and hair.

(Anonymous, 1948–1976)

Roots are expectorant and demulcent. Root bark is saponaceous, and used as detergent. Fruits are emetic, tonic, astringent and anthelmintic, used in asthma, colic due to digestion, diarrhoea, cholera, tuber gland, paralysis of limbs, and lumbago. Pessaries made from kernels are used to stimulate uterus during child-birth. Root bark is used as fish poison. Kernel yield oil which is used for soap making. Wood is used for carts, cabinets, turnery and construction work.

(Ambasta, 1986)

Roots and fruits together are considered as anthelmintic. Fruit are used as intestinal vermicide. Nuts are used as vermifuge for chickens. Stem bark is used on worm infested wounds of calf-laying.

(Jain, 1991)

**Schleichera oleosa* (Lour.) Oken.

Family: Sapindaceae, Local Name: Kusum, Koshimb

Ethnobotanical Notes

Medicine

1. Root extract is externally applied to get relief from body-ache and rheumatism.

Miscellaneous

1. Leaves are used for thatching roofs.
2. Wood is used for house construction.

Notes from Classics

Seed oil is applied on scalp to promote hair growth. It is also applied by natives to cure itch and achne. Bark rubbed up with oil is also used on itch. Bark paste is externally applied to relieve pain in the back and loin. Wood is used to prepare articles like pestles, axles of wheels and teeth of harrows, screw rollers of sugar mills and of cotton and oil presses.

(Watt, 1889–1893)

Stem bark is externally applied on goat, itching, pains in back, loins, inflammations and ulcers. Infusion of bark is prescribed in malaria. Seed oil is used as hair oil, also used to cure itch, acne, burns, and other skin troubles, also used against rheumatism. Seed oil is used for edible purposes and culinary purposes. Timber is used for rice powders, sugarcane, industry and agricultural implements. Wood is used as fuel and for making charcoal.

(Anonymous, 1948–1976)

Bark is used as an external application for itching-pain in back and loin, inflammation and ulcers. Seed oil is applied externally on rheumatism and skin troubles. Young shoots are cooked and eaten. Raw fruits are pickled and ripe fruits are eaten. Wood is used for sugar, oil, and rice mill powders, pestles, axles, hubs, felloes, cart wheels, shaft, mortars, agricultural implements and tool handles. Timber is suitable for construction of cabinet work, wagon building, road paving etc. Wood is used as fuel and makes excellent charcoal.

(Ambasta, 1986)

Bark is applied on gout, itching, paralysis, rheumatism and sprain. Seeds are useful in the treatment of burns, skin diseases and rheumatism. Seed oil is used in wounds, also used as luminant. Seeds are used to apply on veterinary ulcers. Stem bark is used as fish poison. Seed oil is used for cooking purposes. Plant is useful on cholera, pleurisy and pneumonia.

(Jain, 1991)

Critical Notes

Roots are not reported as useful in past.

**Schrebera swietenoides* Roxb.

Family: Oleaceae, Local Name: Mokha

Ethnobotanical Notes

Medicine

1. Stem bark is boiled and decoction is applied in case of joint pains.
2. Leaf juice is dropped in ears against ear-ache.

Miscellaneous

1. Tender leaves are cooked and eaten as vegetable.

Notes from Classics

Wood is used for making cart wheels, weavers beam, combs, mathematical instruments etc.

(Watt, 1889–1893)

Roots are used in the treatment of leprosy. Bark is used as an application for boils and burns. Roots are applied on wounds of cattle to repel worms. Leaves are used in the treatment of enlargement of spleen and urinary discharge. Leaves are eaten in times of scarcity. Timber is useful for making beams, rafters, weavers, looms, shutters, wooden utensils, agricultural implements, carts, and also used as fuel and makes excellent charcoal.

(Anonymous, 1948–1976)

Uses noted by Ambasta (1986) are similar to those recorded by Anonymous (*loc.cit.*).

Roots are used on leprosy. Bark is useful on cough and mental depress. Roots are useful to remove worms from wounds of cattle. Leaves are used on tooth-ache and tooth extraction. Fruits are used on hydrocele.

(Jain, 1991)

Critical Notes

Utilities of stem bark for joint pains and of leaf juice against ear-ache are unreported. Use of tender leaves as vegetable is likewise unnoticed in past.

Securinega leucopyrus (Willd.) Muell.-Arg.

Family: Euphorbiaceae, Local Name: Petuni

Ethnobotanical Notes

Miscellaneous

1. Leaves are cooked and eaten as vegetable.
2. Ripe fruits are edible.
3. Branches are used for fencing.

Notes from Classics

Juice of leaves or paste with tobacco are used to destroy worms in sores. Fruits are eaten. Wood is used as fuel.

(Watt, 1889–1893)

Leaf juice or paste along with tobacco is given to destroy worms in sores. Leaves are eaten. Berries are edible. Slender branches are used for making weak baskets and for thatching huts

(Anonymous, 1948–1976)

Leaf paste or juice along with tobacco is used to destroy worms in sores. Slender branches are used for making huts. Leaves are cooked and eaten. Berries are sweet and edible. Stem bark is used as fish poison.

(Ambasta, 1986)

**Securienega virosa* (Roxb. ex Willd.) Pax & Hoffm.

Family: Euphorbiaceae, Local Name: Pichrundi, Pichurna, Karmati

Ethnobotanical Notes

Medicine

1. Leaf paste is applied on body to get relief from body-ache. This is followed for 2 to 3 days.
2. Leaf paste is applied on dislocated part. It is wrapped for about a week.
3. About two spoons of leaf powder and one spoon of infusion is taken orally at night for three days to treat spermatorrhoea. Fruits are edible.

Notes from Classics

Root is astringent and purgative, used in fever. Roots are reported as analgesic and aphrodisiac. Bark is astringent and used in diarrhoea and pneumonia. Leaf juice or their paste alongwith tobacco is used to destroy worms in sores. Decoction of leaves is used in fever, also used for cleaning wounds. Ripe fruits are eaten. Gum is used as paste. Wood is used for agricultural implements, tent pegs, chair legs, fishing stakes, walking sticks etc. Wood is used as fuel and charcoal.

(Anonymous, 1948–1976)

Roots are analgesic and aphrodisiac. Leaf juice is employed for destroying worms in sores. Decoction of leaves is given as laxative, and antipyretic. Fruits are eaten. Wood is used for agricultural implements, walking sticks, tent pegs, chair legs, fishing roofs.

(Ambasta, 1986)

Leaves are useful in treatment of bone fracture, constipation and worms in sores. Fruits are used as a remedy for stomach-ache.

(Jain, 1991)

Critical Notes

Leaves are important for body-ache and spermatorrhoea. These are new applications of leaves.

**Sesbania bipinosa* W.F. Wight

Family: Papilionaceae, Local Name: Dhedor

Ethnobotanical Notes

Medicine

1. Root paste is applied on goitre to control its further growth.

Notes from Classics

Seeds are eaten in times of famine.

(Watt, 1889–1893)

Plant is used as fodder.

(Anonymous, 1948–1976)

Plant yields fibres used for making ropes and fishing nets. Plant is used as cattle fodder.

(Ambasta, 1986)

Seeds are cooked and eaten.

(Jain, 1991)

Critical Notes

Medicinal value of roots to treat and stop further growth of goitre is yet to be reported.

***Setaria pumila* (Poir.) R. & S.**

Family: Graminae, Local Name: Kohalu

Ethnobotanical Notes

Medicine

1. Awns of inflorescence are collected and crushed in milk. It is then boiled and filtered. One spoonful of this filtrate is given to children to get rid off from tapeworms.

Miscellaneous

1. Plant is eaten by goats in times of scarcity.

Critical Notes

Utility of this species is still not on record.

Sida rhombifolia* L.

Family: Malvaceae, Local Name: Chiknya

Ethnobotanical Notes

Medicine

1. Leaf juice is administered against dysentery in child.

Miscellaneous

1. Branches are used for making brooms for sweeping.
2. Leaf juice is dropped on injury to stop bleeding.
3. Plants are cherished by goats and sheep.

Notes from Classics

Roots are reputed by natives in the treatment of rheumatism. Stem is employed as demulcent and emollient for external and internal use.

(Watt, 1889–1893)

Roots are used as a remedy on rheumatism and leucorrhoea. Stem abounds mucilage is given internally on skin diseases and also considered as diuretic and febrifuge, employed as demulcent and emollient. Plant is recommended to be beneficial in tuberculosis and rheumatism. Plant yields fibre comparable to juice.

(Anonymous, 1948–1976)

Uses recorded by Ambasta (1986) are similar to uses of Anonymous (*loc.cit.*).

Roots are used on asthma, bile, boils, for child-birth, cuts, bruises, gonorrhoea, as heart ailment, used on rheumatism, and is considered as tonic. Leaves are used in dysentery and also applied externally on wounds and swellings. Stem is useful on gum infection, also treated for spermatorrhoea. Entire plant is useful in gonorrhoea, also used as detergent. Stem yield fibres useful for cordage.

(Jain, 1991)

Critical Notes

Utility of leaf juice against dysentery in child is unreported in past.

Using branches for making brooms is unreported yet.

**Smithia conferta* J.E.Sm.

Family: Papilionaceae, Local Name: Kawali

Ethnobotanical Notes

Miscellaneous

1. Leaves are boiled and cooked as vegetable.
2. It is offered to Lord Shiva at Trimbakeshwar.

Notes from Classics

Leaves are eaten as pot-herb.

(Watt, 1889–1893)

Herb is considered to be laxative, and tonic, is given to cure biliousness and rheumatism, also used to cure sterility in women. Leaves are cooked as vegetable.

(Anonymous, 1948–1976)

Uses noted by Ambasta (1986) are similar to the uses reported by Anonymous (*loc.cit.*).

Leaves are used on antifertility. Leaves are edible as vegetable.

(Jain, 1991)

Critical Notes

Food value of the leaves is reiterated by the present documentation. The religious purpose is also probably unknown from other parts of our country.

**Solanum anguivi* Lam.

Family: Solanaceae, Local Name: Ranvange

Ethnobotanical Notes

Medicine

1. Stem fibres are tied using black thread. These are the fastened on ears to treat catarrah.
2. Fruit juice is dropped in ears to treat ear-ache.

Miscellaneous

1. Keeping of stem fibre in abode is thought auspicious. These are collected on the new moon day, which is preferred especially on 'Pola' festival in this region.
2. Leaves are cooked and eaten as vegetable.

Notes from Classics

Roots are ingredient of 'Dasmul', regarded as expectorant, useful in cough, asthma, chronicfebrile affections, colic, flatulence and to cure worms. Fruits are burnt and smoke is a remedy for tooth-ache.

(Watt, 1889–1893)

Roots are expectorant, forming an ingredient of 'Dashmula'. Stem, flowers and fruits are bitter and carminative. Berry juice is used for sore throats. Seeds are expectorant, administered in asthma and cough. Leaf juice and black pepper is used in rheumatism. Fruits are esteemed for curies. Seeds are also eaten. Plant is used as goat fodder.

(Anonymous, 1948–1976)

Roots are carminative and expectorant, used in coughs and cattarahal affections, dysuria and colics. It is pounded and applied on nasal ulcers. Leaves are laxative and digestive. Fruits are laxative and digestive. Half ripe fruits are used in curries and chutneys.

(Ambasta, 1986)

Roots are used on cough and ear-ache. Fruits are used on tooth-ache. Fruits are edible.

(Jain, 1991)

Critical Notes

Stem fibres are thought auspicious on a specific day. They are also useful to treat catarrah. These uses are not noted elsewhere.

**Solanum virginianum* L.

Family: Solanaceae, Local Name: Bhuiringni

Ethnobotanical Notes

Medicine

1. Roots of this plant and fruits of *Borassus flabellifer* Linn. together are crushed and this mixture is given in impotency in male and female. Mixture with honey also increases sexual power. It is administered for nine days.
2. Fruits are cut into two pieces. These are placed on tooth against tooth-ache.

Notes from Classics

Roots have expectorant qualities, given in asthma, cough, catarrhal and pain in chest. Herb is recommended on cough, dysuria, catarrhal fever, leprosy, costiveness, stone in bladder. Seeds are used to cure tooth-ache. Decoction of plant is used as febrifuge and chronic malarious fever.

(Watt, 1889–1893)

Root is expectorant and form ingredient of well known ayurvedic medicine 'Dashmula'. It is used on cough asthma and chest pain. Stem, flowers and fruits are bitter and carminative. They are prescribed for burning sensation of feet accompanied by vesicular watery eruptions. Leaf juice mixed with black pepper is prescribed on rheumatism. Juice of berries is used as a remedy for sore. Plant is diuretic, used to cure dropsy and Dengu fever. Fruits are esteemed for curries. Seeds are eaten after cooling. Herbs are eaten by goats.

(Anonymous, 1948–1976)

Uses noted by Ambasta (1986) are similar to those reported by Anonymous (*loc.cit.*).

Roots are used in the treatment of cold, cough, urine complaint and chest pain. Roots are useful as antidote on snake-bite and scorpion sting. Leaves are used on eye complaints. Fruits are useful on bronchitis alongwith roots used on cough ear-complaint, gum trouble, piles, sores, paralysis, tooth-ache and tooth decay. Seeds are used in migrane, pains in jaws, and they are smoked on tooth-ache. Entire plant is used on asthma, dropsy, fever, gonorrhoea, heart diseases and swellings.

(Jain, 1991)

Critical Notes

Roots are being reported for the first time as useful to treat impotency.

**Solena amplexicaulis* (Lam.) Gandhi

Family: Cucurbitaceae, Local Name: Gommet

Ethnobotanical Notes

Medicine

1. Tubers of this taxon and the tubers of katirle (*Momordica dioica* Roxb. ex Willd.) are crushed together. This paste is chewed, the extract collected in mouth is swallowed, this helps to improve appetite.

Notes from Classics

Roots are used to treat antifertility. Roots are used as tonic, and used on small-pox. Leaves are used on cuts, to treat spermatorrhoea and diabetes. Fruits are eaten as vegetable, useful in fever. Entire plant is useful as cooling, to treat ear-ache, labour, paralysis, and unconsciousness. Roots are used as veterinary medicine on sores. Seeds are used as antidote on snake bite.

(Jain, 1991)

Critical Notes

Tubers are yet unnoted for improving appetite.

Sonchus oleraceus L.

Family: Compositae, Local Name: Mhatara

Ethnobotanical Notes

Miscellaneous

1. Plant is used as cattle fodder.

Notes from Classics

Roots and leaves are used by natives in infusion as tonic and febrifuge. Entire plant is used as galactogogue and in disease of liver. Leaves are cooked as vegetable. Plant is used as cattle fodder.

(Watt, 1889–1893)

Roots are used as vermifuge. Roots and leaves are used in indigestion and as febrifuge. Stems are prescribed as sedative and tonic. An ointment made from decoction of root is used to dress ulcers and wounds. Plant is valued as galactogogue and used in the treatment of ascites and hydrothorax. Plant is cooked as vegetable or salad. Plant is used as sheep and goat fodder.

(Anonymous, 1948–1976)

Roots are used as vermifuge. Roots and leaves are used in indigestion and as a febrifuge. Stem is used as sedative. Gum is said to be cathartic, used for ascites and hydro-thorax. Plant is used as galactogogue and in liver complaints. Herb is consumed as salad and vegetable. Plant is used as cattle fodder.

(Ambasta, 1986)

Leaves are used on cuts and injuries. Stem is used as vegetable.

(Jain, 1991)

Critical Notes

The plant species is exotic, native of Europe.

Sorghum halepense (L.) Pers.

Family: Graminae, Local Name: Barro, Shallu

Ethnobotanical Notes

Miscellaneous

1. Stem and culms are used for thatching huts and for making partition walls.

2. It is used as fodder for cattle.
3. Grains are used for making breads.

Notes from Classics

Grains are eaten by poor classes. Plant is used as cattle fodder. Native pens are made from stems for sheep and goats.

(Watt, 1889–1893)

Sorghum grains are properly cleaned, is an important article food for poor classes. Plant can be fed to livestock in form of green fodder, dry roughage or silage. Grains are used as poultry feed.

(Anonymous, 1948–1976)

Grains are diuretic, demulcent, used as food in times of scarcity. Plant is grown for fodder eaten by cattle.

(Ambasta, 1986)

Aerial roots are used to increase intoxicating effects of drinks.Grains are used on stomach complaints, and also edible as local drink. Entire plant is useful in fever.

(Jain, 1991)

Critical Notes

The uses noted by the present investigators and those mentioned in the literature clearly indicate that the species is very valuable particularly to the tribal people. It grows with little care and usually found in hilly tribal pockets. Plant is exotic, native of Arabia.

Sorghum bicolor (L.) Moench.

Family: Graminae, Local Name: Jawar

Ethnobotanical Notes

Miscellaneous

1. Stem stalks are used for partition walls of huts, which are coated by garbage of wet soil.
2. Dried stem are used as cattle fodder.

Notes from Classics

Millet is an important article of food within the labouring classes. Stalks are called 'Kadbi', a favourite fodder of cattle.

(Watt, 1889–1893)

Seeds are diuretic and demulcent. Grains are ground into flour and made into chapattis. Plant is used as cattle feed. Grains are used as poultry feed and also fed to fattening pigs.

(Anonymous, 1948–1976)

Grains are medicinally used for baldness.

(Jain, 1991)

Critical Notes

It is usually grown as crop in tribal areas. Uses noted by the present investigators and those in classic literature amply indicate that it is one of the important species to support the life of tribals in various ways. Plant is exotic, native of Tropical America.

**Spilanthus calva* DC.

Family: Compositae, Local Name: Akkalkara

Ethnobotanical Notes

Medicine

1. Hairs growing on the floral heads are completely removed and cow-ghee is applied on head against headache.
2. Flower head of this plant and roots of *Indigofera cordifolia* Heyne and *Saccharum officinarum* Linn. (Cane with red colour and very short internodes) all are pounded together. This powder is given with cow-ghee mixed in 1:1 ratio. It is consumed on madness.

Notes from Classics

Flower heads are very acrid it is considered as a powerful stimulant and sialogogue, and is used in head-aches, paralysis of tongue, affections of throat, gums, and tooth-ache. It is employed in India as a remedy for tooth-ache.

(Watt, 1889–1893)

Root is used as purgative. Decoction of plant is given as diuretic, and employed as a bath for rheumatism and as lotion in scabies and psoriasis. Pounded herb is used as poultice to dress wounds. Plant is used as fish poison. Flowers are chewed to relieve tooth-ache, throat and gums and paralysis of tongue. It is a popular remedy for stammering children.

(Anonymous, 1948–1976)

Roots are considered as purgative. Pungent flowers are chewed for relief in throat affection and paralysis of tongue, also used for stammering in children. Tincture of capitula acts as a sialagogue and stimulant and used in caries and inflammation of jaw bones. Herb is used in dysentery. Decoction is given as diuretic and used in scabies and psoriasis. Plant is used as fish poison. Dried capitula contain spilanthol which shows strong sialagogic action, acts as a local anaesthetic, and a powerful insecticide.

(Ambasta, 1986)

Roots are used in cold and cough, also useful in diarrhoea. Leaves are used as anaesthetic, used in tooth-ache and scabies. Entire plant is used in dysentery.

(Jain, 1991)

Critical Notes

Use of floral heads, of course along with other constituents mentioned, are beneficial in madness. This is not documented in the classic literature.

Spondias pinnata (Linn. *f.*) Kurz

Family: Anarardiaceae, Local Name: Ambada

Ethnobotanical Notes

Medicine

1. Fruit juice is given against gastric complaints.

Miscellaneous

1. Wood is used as fire wood.

Notes from Classics

Leaves and bark are astringent and aromatic, administered in dysentery. Leaf juice is locally used on ear-ache. Fruit is described as acid and astringent, useful in bilous dyspepsia. Gum is used as demulcent. Fruits are considered as a antidote for wounds from poisonous arrows and for these purpose eaten either green or dry. Timber is used as fuel.

(Watt, 1889–1893)

Roots are used for regulating menstruation. Bark is useful in dysentery, diarrhoea, and also prevents vomiting. Bark paste is applied for rheumatism. Gum is used as demulcent. Fruit juice is dropped in ears for ear-ache. Fruits are eaten green or dry as antidote for wounds caused by poisonous arrows. Fruits are eaten as vegetable when green and as fruits when ripe, used as condiments and made into chutneys, stews, pickles and jams.

(Anonymous, 1948–1976)

Leaves are stewed or steamed and eaten, sometimes cooked with meat to make tender. Fruits are edible, fresh juice is used for making beverages and sherbets. Unripe fruits are made into jelly, pickles, preserves and marmalades, also used for flavouring soups and stews.

(Ambasta, 1986)

Bark is used against dysentery, stomach-ache and rheumatism. Leaves are used on itching, to cure ring-worms, and also useful in ear-ache. Entire plant is useful to treat cholera, diarrhoea, opthalmia and night fever.

(Jain, 1991)

**Stemodia viscosa* Roxb.

Family: Scrophulariaceae, Local Name: Ramban

Ethnobotanical Notes

Medicine

1. Leaves are burnt and the ash called 'Misri' is applied on white spots detected as preliminary symbol of leprosy.

Notes from Classics

Plant infusion is considered as a demulcent.

(Watt, 1889–1893)

A dried plant is fragrant and used as demulcent.

(Anonymous, 1948–1976)

Dried plants are slightly fragrant and used as demulcent.

(Ambasta, 1986)

Plant is said to be demulcent.

(Jain, 1991)

Critical Notes

Medicinal utility of leaves for treating leprosy is new and noteworthy.

**Sterculia urens* Roxb.

Family: Sterculiaceae, Local Name: Kudal

Ethnobotanical Notes

Miscellaneous

1. Leaves are crushed and extract is thrown in water for stupefying fishes.
2. Seeds after removing seed coats are directly consumed by tribals.

Notes from Classics

Leaves and tender branches steeped in water yield mucilaginous extract employed in pleuro-pneumonia of cattle. Gum is used for throat affections. Seeds are roasted and eaten. Wood is used for making toys and guitars.

(Watt, 1889–1893)

Tender roots are reported to be eaten after cooking. Bark yields fibre used for cordage. Pounded bark is given to women to facilitate delivery. Leaves and tender shoots when steeped in water yield mucilagineous extract used in pleuro-pneumonia in cattle. Seeds are eaten like pulses. Gum is used for printing pastes in textile industry and in adhesives, also used in preparations of lotions, sprays and pastes, used in cosmetics and paper industry. Timber is used for toys, packages, dugouts, match boxes, panels, furniture etc. Wood is used as excellent fuel.

(Anonymous, 1948–1976)

Uses recorded by Ambasta (1986) are similar to uses noted by Anonymous (*loc.cit.*).

Root paste is applied on dislocations and fractures. Stem bark is used in labour. Gum is useful against blood dysentery, joint pains stomach diseases, throat affections, also used as tonic. Entire plant is useful as application on sores. Seeds are edible as vegetable. Gum is used in manufacture of jams and confectionary.

(Jain, 1991)

Critical Notes

Use of leaves as fish poison is yet unknown.

**Streblus asper* Lour.

Family: Moraceae, Local Name: Kharbat

Ethnobotanical Notes

Medicine

1. Roots are fastened near ear as a treatment of cataract.

Notes from Classics

Dry root powder is used on dysentery. Roots are useful externally on ulcers and sinuses. Decoction of stem bark is given in fevers, dysentery and diarrhoea. Juice of plant has astringent and antiseptic qualities and is applied to sores, heels and chapped hands. Roots are used as antidote to snake poison.

(Watt, 1889–1893)

Poultice of roots is applied on ulcers, sinuses, boils, swellings, boils, and pulverized roots are given in dysentery. Decoction of bark is used in fevers, diarrhoea and dysentery. Leaves are used as galactogogue. Infusion of leaves is taken as a substitute for tea. Seeds are used in epistaxis, piles, diarrhoea, and externally their paste is applied on leucoderma. Latex is astringent and antiseptic. It is externally applied to sore heals, chapped hands and glandular swellings, also applied to temples as a sedative in neuralgia. Berries are edible. Leaves are lopped as fodder for cattle and elephants. Leaves are used for polishing wood, ivory, horns, and cleaning utensils. Wood chips are mixed with tobacco, and used in Burmese cheroots, also used for making carts wheels, yokes and agricultural implements. Wood is used as excellent fuel.

(Ambasta, 1986)

Stem is used on tooth-ache. Stem bark is recommended on stomach-ache and urine complaints. Latex is used on swelling chicks. Leaves are used as substitute for polish paper. Entire plant is useful in cancer, cholera, colic, diarrhoea, dysentery, menorrhoea, sores and stomach-ache.

(Jain, 1991)

Critical Notes

Roots are useful to better cataract, a fact still remain unnoticed.

Syzygium cumini (L.) Skeels

Family: Myrtaceae, Local Name: Jamun

Ethnobotanical Notes

Medicine

1. Stem bark of this plant and roots of *Musa paradisiaca* Linn. are crushed together in water. This extract, about 15–20 ml is taken orally every day on yellow urination till cure.
2. Leaves are burnt and the ash called 'Misri' is mixed in coconut oil. It is applied on swelling or inflamed spot called 'Kanduk' on human body. Tribal believed that such swellings or spots appear because of Devil's power.
3. Wine prepared from dried fruits is drunk to avoid cholera.
4. Fruits partly eaten by Owl are collected then decoction is given to children in cases of diarrhoea, gastro and flatulence. Doses vary from 2 to 4 spoons per day till cure.

Miscellaneous

1. 'Rabani' is an agricultural practice for which leafy branches are choffed up from trees which is called as 'Khandani' and the branches are spread all over the field for drying and are burnt for the crop specially Nagali (*Eleusine coracana* Gaertn.) and Varai (*Panicum miliaceum* L.) are sown in next season. The branches of this taxon are particularly used.

Notes from Classics

Bark is astringent, used in case of dysentery. Decoction is used as tooth gargle. Fresh juice of bark is given with goat's milk to children on dysentery, diarrhoea. Fruits are stomachic, carminative and diuretic. Vinegar is prepared from unripe fruits. Wood is used for agricultural implements, carts etc. Leaf juice is astringent and given in dysentery, also used on painful gums. Leaf is used as poultice for scorpion sting. Powdered seeds are given in cases of diabetes.

(Watt, 1889–1893)

Ripe fruits are eaten. A spirituous liquor is distilled from fruits. Fruits are used for making preserves, squashes and jellies. Juice of unripe fruit is used to prepare vinegar. Timber is used for construction, boat building, tea-chest, plywood, agricultural implements, tool handles, cart-wheels, well-curbs, sleepers, furniture and as props for shafts and galleries in mine, also used as good fuel.

(Anonymous, 1948–1976)

Bark is astringent, used in case of dysentery. Bark is used for tanning and dyeing and for colouring fishnets. Infusion of bark used in gargle as mouth wash. Ripe fruits are edible, a spirituous liquor and wine is prepared from fruits. Decoction of bark and seed powder is useful in diabetes. Leaves are used as fodder for sheep, goats and cattle foliage is also used as green manure. Wood is used for construction of houses, boat, plywood, agricultural implements, cart wheels, tool handles, furniture etc.

(Ambasta, 1986)

Stem bark is astringent, used to treat piles. Leaves are chewed for blisters in mouth. Fruits are eaten to cure stomach-ache, also used as tonic. Ripe fruits are eaten. Seeds are useful for diabetic patients and also cure pimples. Entire plant is used in anaemia, cancer, colic, diarrhoea and dysentery. Leaves are used for thatching roofs of huts.

(Jain, 1991)

Critical Notes

Stem bark for treating yellow urination and leaves for curing swellings are new applications.

Tacca leonotopetalides (L.) Oktze.

Family: Taccaceae

Ethnobotanical Notes

Miscellaneous

1. Tubers are cooked or roasted and eaten by tribals.

Notes from Classics

Tuberous roots are eaten by natives.

(Watt, 1889–1893)

Tubers form an important article of food. Starch obtained from tubers is used to prepare porridges, cakes and other sweet meats. It can also be used as laundry starch. Tubers are used in the treatment of piles. Extract of tubers prepared by washing the grated tubers in running water is a rubefacient and also given in diarrhoea and dysentery. Stems, petioles and scapes are used for hat making.

(Anonymous, 1948–1976)

Tubers are useful in the treatment of piles. Extract of tubers is given in diarrhoea and dysentery. Tubers are source of starch. Starch is used in porridges, cakes and sweet meats.

(Ambasta, 1986)

Tubers are used as a remedy for body-ache and head-ache. Tubers are eaten by natives.

(Jain, 1991)

Tamarix ericoides Rottl.

Family: Tamaricaceae, Local Name: Shenwar

Ethnobotanical Notes

Miscellaneous

1. Stems are used to make brooms which are used for sweeping houses and foreground or farmyards.

Notes from Classics

Branches are used for fire wood.

(Watt, 1889–1893)

Decoction of leaves is given in enlargement of spleen. The galls are astringent. Leaves are cooked with rice and given to children to relieve cough. Twigs are employed for brooms and brushes in Maharashtra. They are also used for fire wood.

(Anonymous, 1948–1976)

Leaves are cooked with rice and given to children to relieve cough. Decoction of leaves is given in enlargement of spleen. Twigs are employed for making brooms.

(Ambasta, 1986)

Leaves are used on cough and spleen enlargement. Twigs symbolically drives away ghosts.

(Jain, 1991)

**Tagetes erecta* Linn.

Family: Compositae, Local Name: Zendu, Genda

Ethnobotanical Notes

Medicine

1. Leaf juice two to three drops is poured in ears to relieve ear-ache.

Notes from Classics

Leaf paste is applied on carbuncle. Leaf juice is useful on ear-ache, and acts as sedative. Ointment made from flowers is used on ulcers. Flowers given internally are supposed to purify blood. Juice of petals heated with ghee is a remedy on bleeding piles.

(Watt, 1889–1893)

Roots extract is considered as laxative. Leaves and florets are used as emmenogogoue. This infusion is prescribed as a vermifuge, diuretic and carminative. Leaves are used in the treatment of kidney troubles and muscular pain and also applied on boils and carbuncles. Florets are used in the treatment of eye disease and ulcers. Infusion of plant is used in rheumatism and bronchitis.

(Anonymous, 1948–1976)

Root extract is laxative. Leaves are used in renal troubles, muscular pains, and also applied on boils and carbuncles. Leaves and florets are used as an emmenogogue. Infusion of leaves and florets is prescribed as carminative, diuretic and vermifuge. Florets are helpful for eye complaints. Florets yield yellow dye. Florets yield oil used for chickens, fed to enhance yellow colour of chicken skin and eggs.

(Ambasta, 1986)

Leaves are used on urine complaints, cuts and wounds. Flower juice is dropped in ears for ear-ache.

(Jain, 1991)

Critical Notes

Leaf juice is new application for curing ear-ache. It is cultivated for its flowering heads. Plant species is exotic, native of Mexico.

Tamarindus indica L.

Family: Caesalpiniaceae, Local Name: Chinch, Imali

Ethnobotanical Notes

Medicine

1. Leaf juice is administered to improve mouth taste after fever.

Veterinary Medicine

1. Leaves of this plant and soil of ant-hill are mixed in water. It is boiled and then applied on swellings in case of cattle.

Miscellaneous

1. Ripe and unripe fruits are consumed by tribals.

Notes from Classics

Fruit pulp is used as refrigerant, carminative and laxative, and prescribed in febrile diseases, bilious disorders, also used in the form of infusion. Tamarind infusion in recommended for sore throat and as antiseptic. Seeds pounded with water are used on cough and relaxation of uvula. Poultice of fruits is used on swellings. Leaves are eaten by goats as a fodder. Kernels are used in varnishes, and for burning lamps. Fruit pulp is used as agent for souring curries, sauces, chutneys and few beverages. Seeds are roasted and eaten. Ground seeds are used as cattle feed.

(Anonymous, 1948–1976)

Infusion of fruits is given on biliousness and febrial condition. Fruit pulp is laxative, carminative and refrigerant. Leaves and seedlings are eaten as vegetable. Leaves are used as fodder and also as green manure. Kernels are used in times of scarcity. Fruits are used for making jams, jellies and marmalades. Seed oil is used in paints and varnishes. Wood is used for constructions agricultural implements pounders.

(Ambasta, 1986)

Leaves are used to treat eye complaint, fever, stomach-ache, urine complaints and as antihelmentic. Seeds are used on boils, and also used as soft drink. Plant is useful in anasarca, dropsy, head-ache, muscle pain and syphilis. Plant is used as antidote on snake bite and scorpion sting. Leaves and flowers are eaten as vegetable. Fruits are used in curries, chutneys and also improves digestion.

(Jain, 1991)

Critical Notes

Plant species is exotic, native of Tropical Africa.

**Tamilnadia uliginosa* (Retz.) Trivengadum & Sastre.

Family: Rubiaceae, Local Name: Pendar

Ethnobotanical Notes

Medicine

1. Stem bark powder is applied on septics and deep injuries of cattle.

Notes from Classics

Wood is recommended for making boxes and combs.

(Watt, 1889–1893)

Timber is used for making boxes, combs, turnery, articles, light furniture, tobacco jars, shuttles, toys, mathematical instruments etc.

(Anonymous, 1948–1976)

Ambasta (1986) mentioned uses similar to those in Anonymous (*loc.cit.*)

Stem is used to treat, sores of hands and feet. Fruits are used on stomach-ache. Leaves are used to cure veterinary wounds. Fruits are edible.

(Jain, 1991)

Critical Notes

The ethnoveterinary use noted presently of the stem bark is additional.

**Tectona grandis* L.*f.*

Family: Verbenaceae, Local Name: Sag

Ethnobotanical Notes

Medicine

1. Roots of one year seedling are tied in ear, after applying red lead and 'Dhup' (Frank-incense) as a treatment of cataract.
2. Leaf petiole is chewed and tied near urinogenital opening around loin as a remedy for burning sensation in urination.
3. Single seed is chewed daily for 8 to 10 days as a remedy for kidney stone.
4. Seeds of these plant are finely crushed with seeds of Kadam [*Mitragyna parvifolia* (Roxb.) Korth.] ghee is added in it. It is consumed by a person suffering from kidney complaints.

Miscellaneous

1. Leaves are used as a covering for grains stored in storage tank, also used to cover the hiccups of crops is farm yards.

2. A symbol worshiped by tribals "Wagh Dev" is carved on a log of wood. A long pole of 80–100 feet with a white flag is hoisted at "Diwali festival". Wagh Dev is also worshiped at Diwali festival.
3. Wood is used for hut construction, agricultural tools, carts, toys, kitchen wares, musical instruments and also used for making charcoal.

Notes from Classics

Bark powder is used as tonic, considered as astringent, and used as a remedy for leucorrhoea. Wood powder is plastered on head for head-ache, and for dispersion of inflammatory swellings, dyspepsia, with burning pain in stomach arising from over flow of bile, also used as vermifuge. Oil from nuts is used for hair growth and skin diseases. Wood is used in ships, sleepers, bridges, furniture, houses, railway carriages etc.

(Watt, 1889–1893)

Powder of teak wood is useful as an application for skin inflammation caused by melanorrhoea, when taken internally said to relieve burning sensation in stomach. Wood wash is applied on swollen eyelids. Wood oil is used on eczema and ring-worm. Wood-tar is applied to prevent maggots-breeding in sores in cattle. Timber is used for ship building, bridges, poles, in harbour, railway coaches, furniture, lorry carriages, agricultural, musical instruments and used for all sorts of articles.

(Anonymous, 1948–1976)

Stem bark is astringent and used in bronchitis. Flowers are used on biliousness, and urinary discharges. Flowers and seeds are considered as diuretic. Wood oil is applied on eczema. Root bark is used for colouring mattings. Leaves are used for thatching. Wood is used in all types of constructions of houses, ship, railway, musical instruments, plywood etc.

(Ambasta, 1986)

Stem bark is used on stomach-ache and diarrhoea. Wood oil is used on eczema, eye diseases, head-ache, ring-worm and swellings. Seeds are used in urinary complaints. Wood oil is used on veterinary sores. Wood is used for musical instruments. Leaves are used as umbrellas. Leaf juice is used as red-dye for application on furniture.

(Jain, 1991)

Critical Notes

Roots and petioles are not noted in past as useful for cataract. Seeds although reported beneficial in urinary complaints, but are not expressly noted in case of kidney-stone.

**Terminalia arjuna* (Roxb. ex DC.) Wight & Arn.

Family: Combretaceae, Local Name: Arjun Sadada

Ethnobotanical Notes

Medicine

1. Stem bark is chewed by person having 'White spots' on skin.
2. Ash of stem bark mixed in oil is applied on skin burning.
3. Stem bark of this plant and stem bark of 'Sabar' (*Euphorbia neriifolia*) are burnt and this ash is mixed with coconut oil. It is applied on burn injuries.

Notes from Classics

Stem bark is astringent and cooling, used in heart diseases, fractures, ulcers etc. Leaf juice is poured in ear for ear-ache. Wood is used for house building and cart wheels. Stem bark is used for tanning.

(Watt, 1889–1893)

Bark is acrid, tonic, febrifuge and antidysenteric and also used on fractures. Decoction of bark is used as wash on ulcers. Leaf juice is dropped in ears for ear-ache. Timber is used for house construction, agricultural implements, carts, plywood, tea chests etc. Bark is used for tanning.

(Anonymous, 1948–1976)

Stem bark is styptic, tonic, febrifuge and antidysenteric. Pulverized bark gives relief in symptomatic hypertension, and acts as diuretic in cirrhosis of liver. Fruits are tonic and deobstruent. Leaf juice is used in ear-ache. Stem bark is used for tanning. Wood is used for carts and carriages, agricultural implements, water troughs, boat building. Plywood, house building, tool handles, jetty piles etc.

(Ambasta, 1986)

Bark is used to treat dysentery and heart diseases. Leaf paste is applied on wounds. Plant is used in dysuria, leprosy, neural pain, pleurisy, pneumonia and sores. Stem bark is used as fish poison.

(Jain, 1991)

Critical Notes

Uses of stem and bark for skin diseases and burning are new reports.

**Terminalia bellirica* (Gaertn.) Roxb.

Family: Combretaceae, Local Name: Beheda

Ethnobotanical Notes

Medicine

1. Infusion of stem bark, 2 to 3 spoons, is useful in the treatment of intestinal worms in children.
2. Stem bark juice, about two spoons, is given twice a day for 2 to 3 days as treatment of cholera.
3. Decoction of fruits, one cup twice a day for 2 top 3 days, is taken orally to improve mouth taste especially after illness.
4. One spoon fruit powder and a pinch of *Curcuma domestica* Valeton powder are given twice for 10–15 days to treat chronic cough.
5. Fine dried stem pieces are taken and red lead is sprinkled on them giving a frank incense of censer (Dhup). An incantation is murmured, the stem pieces are tied around loin, or waist for treating cramps and aching.

Veterinary Medicine

1. Fruits of this plant and *Chrysopogon fulvus* (Spr.) Choiv. are crushed in water. This extract is given to hens to drink against 'Fapat' disease of hens.

Miscellaneous

1. Dye extracted from fruits is used by charmers for colouring leather.

Notes from Classics

Fruits are considered as astringent and laxative, useful on cough, hoarseness, eye diseases, also used on dropsy piles, diarrhoea and leprosy. Leaves are fed for milking cows. Fruits are eaten by goats, sheep, cattle, deer and monkeys. Wood is used for house construction, cart-wheels and agricultural implements.

(Watt, 1889–1893)

Fruit pulp is used in dropsy, piles, diarrhoea and leprosy, also used on fever. Half ripe fruits are used on purgative, pulp mixed with honey is used on opthalmia. Ripe fruits are astringent. Fruit oil mixed with ghee is useful on rheumatism, also applied on hair. Kernel oil is used in manufacturing soaps. Timber is used for carts, house building rafters and packing cases.

(Anonymous, 1948–1976)

Stem bark is diuretic. Ripe fruits are astringent, also used as purgative. Fruit oil is applied on hair, also applied to treat rheumatism and swellings. Oil mixed with honey is used on opthalmia. Kernel oil is edible. It is used in soap making. Wood is used for carts, rafters, packing cases, agricultural implements and catarmans.

(Ambasta, 1986)

Stem bark is used on fever and cold. Stem exudates is used on itch by ticks. Fruits are used on bronchitis, constipation, cough, diarrhoea, dropsy, intestinal complaints, leprosy, liver complaint, piles, respiratory diseases, and are considered as purgative and laxative. Plant is used to treat cholera, cramps, dysuria, muscle pain, fever. Plant is used as antidote on snake-bite. Seeds are used on gastric complaints. Fruits are used on vegetable.

(Jain, 1991)

Critical Notes

Applications of stem bark for cholera, fruits to improve taste and to treat cough are yet to be reported.

**Terminalia chebula* Retz.

Family: Combretaceae, Local Name: Hirda

Ethnobotanical Notes

Medicine

1. Seed extract is given to children against cough.

Miscellaneous

1. Wood is used for agricultural implements.
2. Leaves are also used as fodder for cattle.

Notes from Classics

Fruits are used as digestive, fine paste of fruits is applied on burns and scalds. Infusion of fruits is cooling to eyes. Fruits are chewed to tighten teeth. Fruit powder is smoked to get relief from asthma. Ripe fruits are used on fever, cough, urinary diseases, piles, intestinal worms, chronic diarrhoea, constipation, costiveness, flatulence, vomiting, hiccup, heart diseases, enlarged spleen, liver, ascites and skin diseases. Fruit is purgative, removes bile phlegm, adust bile and considered as astringent, used in dysentery and diarrhoea. Leaves are eaten as fodder by cattle. Wood is used for furniture making, agricultural implements, houses, carts, sleepers etc.

(Watt, 1889–1893)

Fruit pulp is used as dentrifice and to cure ulceration from gums. Infusion of fruits is used as a cooling eye wash. Powder is smoked for relief from asthma. Fine powder with common oil is used for burns and scalds. Fruit extract is used in various process such as petroleum purification, cement manufacture, colouring slate stones.

(Anonymous, 1948–1976)

Stem bark is diuretic and cardio tonic. Fruits are laxative, stomachic, tonic, alterative form of 'triphala churna'. Fruit pulp is used in dentrifices. Fruit powder is smoked in asthma. Dried flesh of fruit is an important tanning material.

(Ambasta, 1986)

Bark is used on eczema. Fruits are used in diabetes, indigestion, dysentery, menstrual complaints, pneumonia, eye diseases, sore throat and stomach complaints. Fruits are said to be purgative. Plant is used in bronchitis, cough, constipation, dysuria, measles and sores. Seeds are edible.

(Jain, 1991)

Critical Notes

Seeds are not reported earlier useful against cough.

**Terminalia crenulata* Roth

Family: Combretaceae, Local Name: Sadada

Ethnobotanical Notes

Miscellaneous

1. Leaves are employed to thatch the stalls on roads at the time of circumambulation of 'Bramhagiri' mountain at Trimbakeshwar on the third Monday in month of Shrawan.
2. Large wooden dog of 10′ to 12′ by 11″ to 12″ is carved out in rectangular shape. It is pulled by bullocks to break large soil clods in field formed after ploughing. It is called "Phali Ghalne" in this district.
3. Branches are used for 'Rabani' and agricultural practice before sowing 'Nagali' and 'Warai' seeds.

Notes from Classics

Decoction of bark is given as a remedy for atonic diarrhoea and also used as an application to callous ulcers. Ash of bark are eaten by natives as a substitute for lime with betel-leaf. Leaves are lopped for cattle fodder. Bark is used for poisoning fish by Bhils. Wood is used for fuel and charcoal making.

(Watt, 1889–1893)

Wood is useful for building and agricultural purposes.

(Anonymous, 1948–1976)

Bark is styptic and cardiotonic. Gum is purgative and also used as adhesive. Stem bark is used for dyeing and tanning. Leaves are used for feeding Tasar silk worms. Wood is used for beams, rafters, house constructions, furniture, pounders, carpentry, and agricultural implements, plywood manufacture, manufacture of unwrapping paper, also used in fire proof buildings.

(Ambasta, 1986)

Stem bark is diuretic, used as a remedy in diarrhoea and vomiting. Gum is edible.

(Jain, 1991)

Critical Notes

The miscellaneous uses are as per the needs of the local people, although yet unreported.

Thespesia lampas (Cav.) Dalz. & Gibs.

Family: Malvaceae, Local Name: Ranbhendi

Ethnobotanical Notes

Medicine

1. Roots are chewed against yellow urination. This is repeated for seven days twice a day. Drinking tea, use of spicy food or pungent foods are prohibited during these days.

Miscellaneous

1. Wood is used for making yoke used to keep on shoulders of bullocks to pull the carrier or cart.

Notes from Classics

Roots and fruits are used as a remedy in gonorhoea and syphilis

(Watt, 1889–1893)

Roots are employed in the treatment of gonorrhoea and syphilis. Stem yield fibre used for fabrics, cordage and fishing nets. Timber is used for making drums and other frames.

(Anonymous, 1948–1976)

Roots and fruits are employed for gonorrhoea and syphilis. Flowers are used as applications for cutaneous diseases. Wood is used for making drums and frames.

(Ambasta, 1986)

Roots are used in stomatitis, urine complaints and against sunstroke. Stem is used as a remedy for acidity. Leaves are considered as anti-inflammatory and used on gonorrhoea. Plant is used against bleeding of nose, bronchitis, carbuncle, cough, dysentery, fever, sores, ulcers and wounds.

(Jain, 1991)

**Tinospora cordifolia* (Willd.) Miers ex Hook. *f.* & Thoms.

Family: Menispermaceae, Local Name: Gulvel, Ghamoli

Ethnobotanical Notes

Medicine

1. One cup of decoction of stem pieces is daily given for five to six days to a person suffering from typhoid.
2. A small twig is collected early in the morning, and 'Shendur' (red lead) is applied on it. It is tied with black thread around the neck for seven days to treat jaundice.
3. Stem and leaves are crushed and infusion is given to a person of snake bite.

Notes from Classics

Watery extract of root is a remedy on leprosy and fever. Leaves are used as good fodder for cattle. Decoction of entire plant is regarded as alterative and tonic, used in debility, fever, jaundice, skin diseases, rheumatism, urinary diseases, dyspepsia and secondary syphilis.

(Watt, 1889–1893)

Root is a powerful emetic and used for visceral obstructions. Its watery extract is used in leprosy. Decoction of leaves is used for the treatment of gouts and young leaves bruised in milk are used as liniment in erysipelas. Leaves are beaten with honey and applied on ulcers. Dried and powdered fruits mixed with honey or ghee is used as a tonic and also in the treatment of jaundice and rheumatism.

(Anonymous, 1948–1976)

Roots are emetic and used for visceral obstruction. It sweeter extract used in leprosy. Stem is used in general debility, dyspepsia, fevers and urinary diseases. Stem is used as tonic. Stem bark is antispasmodic, antipyretic, and inflammatory.

(Ambasta, 1986)

Stem is used as a remedy for asthma, cough, jaundice, to increase sex strength, skin diseases, stomach diseases, malaria and diptheria. Leaves are antipyretic, used to treat fever, headache and piles. Seeds are used on pulmonary tuberculosis. Stem is used as veterinary medicine.

(Jain, 1991)

Critical Notes

Stem is being reported for the first time to treat jaundice. Stem and leaves are not reported earliest against snake bite.

**Trewia polycarpa* Benth. ex Hook. *f.*

Family: Euphorbiaceae, Local Name: Petari

Ethnobotanical Notes

Medicine

1. Leaves are dried in sun and then burnt. The ash called 'Misri' is applied as a remedy on piles.

Miscellaneous

1. Ripe fruits are eaten by tribals.
2. Gum is eaten, mixed in sweet-balls (ladu).

Notes from Classics

Fruits are edible. Wood is used for making agricultural implements, packing cases, match boxes, dugouts, yokes, drunks, planks, frames and toys.

(Ambasta, 1986)

Bark is used on stomach-ache. Entire plant is useful to treat swelling.

(Jain, 1991)

Critical Notes

Leaves as a remedy against piles and edibility of gum are still unreported.

**Trichodesma indicum* (L.) Lehm.

Family: Boraginaceae, Local Name: Aghirda

Ethnobotanical Notes

Medicine

1. Ash of dried leaves, locally called 'Misri' is mixed with coconut oil and is applied on injuries till cure.

Notes from Classics

Root paste is applied on swelling of joints. Leaves are used as emollient poultice. Plant is used as antidote on snake-bite.

(Watt, 1889–1893)

Root paste is applied on swelling of joints. Root powder is used on dysentery and fever. Plant is used as emollient poultices. Leaves and flowers are edible.

(Anonymous, 1948–1976)

Roots are used on dysentery. Root powder is applied on swellings. Infusion of leaves is depurative. Herb is emollient, diuretic, and prescribed for expulsion of dead foetus. Leaves and flowers are edible.

(Ambasta, 1986)

Roots are used on stomach-ache. Roots are used as veterinary medicine. Leaves are used on eczema. Entire plant is used on boils, and also as brain tonic. Fruits are used on swellings.

(Jain, 1991)

Critical Notes

Leaves are being reported useful for treating injuries for the first time.

**Tricholepsis indica* (Willd.) Almeida

Family: Compositae, Local Name: Bramha-dandi

Ethnobotanical Notes

Miscellaneous

1. Leaves are used as vegetable.

Notes from Classics

Plant is a nervine tonic and aphrodisiac.

(Watt, 1889–1893)

Root bark is employed in urinary troubles and cough. Herb possess antiseptic properties, and is employed in skin diseases. It is a nervine tonic and given in seminal debility.

(Anonymous, 1948–1976)

Root bark is antiseptic, employed on skin troubles, and also as nervine tonic, used in seminal debility.

(Ambasta, 1986)

Entire plant is used on leprosy mennorrhoea and skin diseases. Roots are used as antidote on snake bite.

(Jain, 1991)

Critical Notes

Edibility of leaves is not noted in past.

**Trichosanthes tricuspidata* Lour.

Family: Cucurbitaceae, Local Name: Kawalyacha-amba

Ethnobotanical Notes

Medicine

1. Stem and fruits are crushed in water. This extract is given to person of snake-bite. It results in vomiting and reduces poison.

Notes from Classics

Root paste is applied to carbuncle, combined with three

myrobalons and turmeric in infusion form is given on gonorrhoea. Roots are described as useful in inflammation of lungs in cattle.

(Watt, 1889–1893)

Root paste is used in carbuncles. Fruits are pounded and mixed in coconut oil for application on sores. Root is also used as veterinary medicine.

(Anonymous, 1948–1976)

Root paste is used on carbuncles. Fruits pounded in oil is applied on sores.

(Ambasta, 1986)

Roots are used in diarrhoea, fever, pneumonia, rheumatism. Leaves are burnt on burns. Seeds are used on dysentery. Fruits are edible.

(Jain, 1991)

Critical Notes

Stem and fruits are helpful to reduce poison. These are not so recorded in past.

**Tridax procumbens* L.

Family: Compositae, Local Name: Tantni, Dagadphodi

Ethnobotanical Notes

Medicine

1. Fresh leaf juice is applied on fresh cut to avoid infection.
2. Leaf juice is dropped into ears of opposite side of the place of the scorpion sting.

Notes from Classics

Leaf juice possess antiseptic, insecticidal, parasiticidal properties, used to check haemorrage, for cuts, bruises and wounds. Leaves are employed in bronchial catarrah, dysentery, diarrhoea and for restoring hairs. Leaves are cooked and eaten as vegetable. Leaves are eaten to as vegetable.

(Anonymous, 1948–1976)

Leaves are useful in the treatment of bronchial cattarah, dysentery and diarrhoea. Leaf juice is used to check hemmorage and wounds. Leaf juice is considered as insecticidal and pesticidal.

(Ambasta, 1986)

Roots are used as remedy for fever. Combination of leaf and stem is used for stone in urinary bladder. Leaves are antiseptic, used on blisters, boils, cuts and wounds, eczema, dysentery diarrhoea, eye diseases and leprosy, skin diseases, sores and stomach-ache and tooth-ache.

(Jain, 1991)

Critical Notes

Leaf juice is employed as antidote for scorpion sting. It is unreported yet. The plant species is exotic, native of South America.

**Triumfetta rhomboidea* Jacq.

Family: Tiliaceae, Local Name: Chiknya

Ethnobotanical Notes

Medicine

1. Roots are crushed and extract is given twice a day for three days to control dysentery and stomach-ache.

Miscellaneous

1. Tender leaves are cooked and eaten as vegetable.

Notes from Classics

Plant is used as demulcent. Mucilage is injected for invertebrate gonorrhoea. Plant is eaten as pot-herb.

(Watt, 1889–1893)

Root is used in dysentery and the bark and fresh leaves are used in diarrhoea. Pounded roots are given for the treatment of ulcers and their hot infusion is said to hasten parturition. Leaves and flowers are used in leprosy. Plant is used as pot-herb in times of scarcity. Stem fibres are stripped off and ribbons are used for ropes and fishing nets.

(Anonymous, 1948–1976)

Roots are pounded and given for intestinal ulcers, also used to control dysentery. Leaves and bark are used in diarrhoea. Leaves and flowers are used in leprosy. Stem is scrapped into ribbons, used for hunting or fishing nets and ropes. Bark yields fibre substitute for jute.

(Ambasta, 1986)

Root mucilage is used in child-birth. Leaves are used in diarrhoea, dysentery, impotency and stomach-ache. Leaf and bark together are used on dysentery. Plant is useful to treat carbuncle, sores, whitlow, and on antifertility. Leaves are edible.

(Jain, 1991)

**Tylophora dalzelli* Hook.*f.*

Family: Asclepiadaceae, Local Name: Kawalvel

Ethnobotanical Notes

Medicine

1. Stem bark of this plant and fruits of 'Bel' (*Aegle mamelos* ex Roxb.) and kernels of 'Coconut' (*Coccus nucifera* L.) are thoroughly mixed. This mixture is wrapped is leaves of *Philiostigma malabaricum* (Roxb.) Benth. and bidis are prepared. It is smoked to treat throat sores and infection. This is followed till cure.

Notes from Classics

Leaves are used in the treatment of asthma.

(Jain, 1991)

Critical Notes

Bark is beneficial to treat throat sores, a fact still remained unnoticed.

**Typha australis* Schum. & Thonn.

Family: Typhaceae, Local Name: Ramban

Ethnobotanical Notes

Medicine

1. Inflorescence is applied on wounds to stop bleeding.

Miscellaneous

1. Leaves are used to prepare trays and pans and used for agricultural purposes.

Notes from Classics

Spadix are applied like cotton to wounds and ulcers. Young shoots are edible. Plant is a favourite fodder for elephants.

(Watt, 1889–1893)

Rhizome is considered as astringent and diuretic. Young shoots, inflorescence and rhizomes are eaten in various ways. Ash of flower spikes is used to heal wounds. Leaves are plaited into several articles such as winnowing trays mats, ropes etc. Silky florets are used for stuffing.

(Anonymous, 1948–1976)

Rhizomes are astringent and diuretic. Rhizomes, young shoots and inflorescence are eaten in various ways. Flowers are made into sweetmeat. Leaves are plaited in ropes, winnowing trays, mats etc.

(Ambasta, 1986)

Fruit extract stops bleeding. Plant is used in epilepsy and madness. Entire plant is useful for thatching roofs of huts, making brooms and mats.

(Jain, 1991)

Critical Notes

To stop bleeding inflorescence is a new application.

**Urena lobata* L.

Family: Malvaceae, Local Name: Waghach-panja

Ethnobotanical Notes

Medicine

1. Roots are chewed on dysentery.

Miscellaneous

1. Stem piece is kept in pocket or under cap. It is believed that tigers do not attack when you are in forest.
2. Fruits are cooked and eaten by natives.

Notes from Classics

Root are used externally as a remedy for rheumatism.

(Watt, 1889–1893)

Roots are diuretic. Poultice made from roots and leaves is used as emollient. An infusion of flowers is used as gargle for apthae and sore throat. Plant yields fibre, used for sacks, ropes, cordage and carpets.

(Anonymous, 1948–1976)

Ambasta (1986) recorded similar uses as those recorded by Anonymous (*loc.cit.*).

Root and stem bark are used in treatment of hydrophobia. Roots are used on lumbago. Stem is used as fibre for miscellaneous use.

(Jain, 1991)

Critical Notes

All uses noted by the present investigators are new. Plant is exotic, probably native of Africa.

**Ventilago denticulata* Roxb.

Family: Rhamnaceae, Local Name: Pivalvel

Ethnobotanical Notes

Medicine

1. Stem bark of this plant, bark of *Syzygium cuminii* (Linn.) Skeels, *Ceiba pentandra* (Linn.) Gaertn. and roots of *Chrysopogon fulvus* (Spr.) Chiov. Roots are crushed together this extract about 15-20 ml twice a day is administered orally to treat dysentery till cure.

Notes from Classics

Baskets are prepared from long stems by Santals.

(Watt, 1889–1893)

Stem is used for basket making. Leaves are used in the treatment of bilious hematuric fevers in Africa.

(Anonymous, 1948–1976)

Pulverized bark is applied on sprains. Sap is used as a remedy for deafness. Seed oil is used for cooking purposes. Bark yields fibre used for cordage.

(Ambasta, 1986)

Roots are used on ear-ache or ear discharges. Bark is cooling, used on rheumatism, spermatorrhoea sprains and stomach ulcers. Stem is used on night blindness. Seeds are abortifacient. Entire plant is useful in dysuria, fever, hair fall, retention urine, faeces, scabies, urine complaints and syphilis. Roots are used as antidote on snake bite. Bark yields fibre for cordage.

(Jain, 1991)

Critical Notes

Stem bark is beneficial in the treatment of dysentery. It is not reported earlier.

Vicoa indica DC.

Family: Compositae, Local Name: Bhangser

Ethnobotanical Notes

Miscellaneous

1. A branch of this plant is selected for worshiping 'Dongri Dev' for Bhagats. On this stem piece and a plate of 'Kanse' (A metal amalgam of Zinc and Copper). Wax is applied. Stem piece is placed vertical on plate and pulled by both hands. It makes vibrating sound. This is done to entertain the 'Dongri Dev'. Simultaneously, a religious tale (Katha) is told.

Notes from Classics

Leaves are used as cattle fodder.

(Ambasta, 1986)

Roots are used on antifertility, tooth-ache and gum complaints. Leaves and roots together are used on head-ache and stomach complaints. Twigs are used as abortifacient.

(Jain, 1991)

Critical Notes

The plant finds place in the worship of a local deity.

Vigna radiata (L.) Wilczek.

Family: Papilionaceae, Local Name: Udid

Ethnobotanical Notes

Miscellaneous

1. Plant is used as cattle fodder.
2. Seeds are cooked and eaten.

Notes from Classics

Decoction of seeds is diuretic, given in beri-beri. Tender pods are consumed as vegetable. Seeds are parched, made into porridge often fried and used as snacks. Plant is fed to cattle after harvesting pods.

(Ambasta, 1986)

Seeds are cooked and consumed.

(Jain, 1991)

Critical Notes

It is cultivated pulse crop. Its stalk and foliage are used as fodder for pet animals.

**Vigna trilobata* (Linn.) Verdcourt

Family: Papilionaceae, Local Name: Mirchi-kand

Ethnobotanical Notes

Medicine

1. Root-stock is chewed and the extract is swallowed to treat stomach-ache and cough.
2. Tubers are chewed as antidote against snake bite. It results in vomiting and reduces poison.

Notes from Classics

Leaves are used as tonic and sedative, used in form of cataplasm for weak eyes. Decoction of leaves is given on irregular fever. Roots are eaten as arrow-root in China. Pulse is consumed by poorer class people. Plant is employed as a cattle feed.

(Watt, 1889–1893)

Anonymous (1948–1976), Ambasta (1986) have recorded uses similar those recorded by Watt (*loc.cit.*).

Fruits are consumed as vegetable. Seeds are cooked and consumed.

(Jain, 1991)

Critical Notes

Roots help to better stomach-ache and cough. Tubers are used as antidote on snake-bite. Both uses are new.

**Vitex negundo* L.

Family: Verbenaceae, Local Name: Nirgudi

Ethnobotanical Notes

Medicine

1. Five pieces of each roots of this plant and those of *Piliostigma malabaricum* (Roxb.) Bth. and *Pogostemon parviflorus* Bth. are collected. These are fastened on ears against tooth-ache.
2. Leaves are boiled and paste is applied on joints for three days against joint pains due to rheumatism.
3. Leaf juice, 2 to 3 drops, are poured in eye against conjunctivitis till cure.
4. Leaves of this plant and of *Dalbergia lanceolaria* L.*f.* are rubbed on hands. Juice obtained is applied on body of a person suffering from fever.
5. Fresh twigs are passed over the body from head to legs as a remedy on fever.

Miscellaneous

1. Leaves are used for thatching roofs of huts.
2. Dried branches are used as brooms.
3. Wood is used as fire wood.
4. Wood is used for making tool handles.

Notes from Classics

Roots are considered as tonic, febrifuge and expectorant. Leaves are aromatic, tonic and vermifuge. Decoction of leaves is given in catarrhal fever and also used on rheumatism. Leaf juice is said to remove foetid discharges and worms from ulcers. Leaf oil is applied to sinuses and scrofulous sores. Leaves are employed to form a warm-bath for women after delivery. Leaf powder given with water cures intermittent fevers. Dried fruits are considered as vermifuge.

(Watt, 1889–1893)

Roots posses tonic, febrifugal, expectorant and diuretic properties, used in dyspepsia, rheumatism and for boils. Root powder is anthelmintic and used as a demulcent in dysentery. Leaves are considered tonic and vermifuge. Decoction of leaves with addition of (*Piper longum*) is given in catarrhal fever, the vapours of leaf decoction are employed in bath for rheumatic affections. Juice of leaves is used in the treatment of foetid discharges. Ointment made from juice is used as hair-tonic. Flowers are astringent; used in fever, diarrhoea and liver complaints.

(Anonymous, 1948–1976)

Roots are anthelmintic, also employed as demulcent, in dysentery and piles. Roots are used as tonic, febrifuge, diuretic, used in rheumatism and dyspepsia. Leaves are considered as tonic and vermifuge. Leaves are smoked for relief in cattarah and head-ache. Leaf decoction is employed in medicinal bath for catarrhal and rheumatic affection. Leaf extract in having anticancer activity. Flowers are astringent, used in diarrhoea, fever and liver complaints. Seeds are eaten after boiling. Young shoots are employed for basketery.

(Ambasta, 1986)

Roots and fruits together are used on diarrhoea, considered as diuretic and expectorant. Stem is used on bone fracture and to treat fever. Leaves are used as applications on blisters, boils, body-ache, cold, cough, fever, gum trouble, head-ache, itching, paralysis, piles, rheumatism, scabies, swellings, ulcers, tooth-ache, wounds, and also reduces sex desire, used as tonic and vermifuge. Flowers are used on dysentery, syphilis, eczema and other skin diseases. Seeds are used on gout. Entire plant is useful in dropsy, eye complaint, mental diseases, post-natal sores and epilepsy. Twigs are used to make basketry also, used as insect repellent.

(Jain, 1991)

Critical Notes

Applications of roots and leaves to treat tooth ache and conjunctivitis respectively are new.

**Wattakaka volubilis* (L.*f.*) Stapf.

Family: Asclepiadaceae, Local Name: Bokudvel

Ethnobotanical Notes

Miscellaneous

1. Fruits and leaves are used as goat feed.

Notes from Classics

Roots and tender stalks are considered emetic and expectorant. Leaves are applied to boils and abscesses. Leaves are used as vegetable.

(Watt, 1889–1893)

Roots and tender stalks are purgative and emetic. Leaves are used in applications for boils and abscesses. Leaves, flowers and rind of unripe fruits are boiled and eaten as vegetable, also used in curries. Seeds are edible.

(Ambasta, 1986)

Roots are anthelmintic and used in fever and regurgitation. Leaves are used on boils, wounds, abscesses and skin diseases. Whole plant is used on anasarca, dropsy, failure of lactation, sore and uriticaria. Leaves and roots are used against gastric complaints in cattle. Plant is used as antidote on snake-bite.

(Jain, 1991)

Critical Notes

Food value is already documented. Fodder value for goats can be easily conceived although not noted in past.

**Woodfordia fruticosa* (L.) Kurz

Family: Lythraceae, Local Name: Dhaiti, Dhait

Ethnobotanical Notes

Medicine

1. Stem bark is crushed in water and this extract, about 25 ml, is given to mothers in case of excessive bleeding after delivery for 6–7 days.
2. Half cup of decoction of stem bark given to control bleeding during dysentery.
3. Flowers of this plant and those of *Butea monosperma* (Lam.) Taub. are homogenized with jaggery which is five year old. Twenty one tablets. are prepared and given thrice a day for conception.
4. Leaves are used as antidote against snake bite.

Notes from Classics

Two drachms of dried flowers are given with curdled milk, in dysentery and with honey in menorrhagia. Flower powder is used on ulcers. Flower are also prescribed in pregnancy, bilious and mucous disorders, and haemorrhoids. Decoction of flowers is given in female complaints. Fruits are used as stimulant and astringent, used in bowel complaints and haemorrhages.

(Watt, 1889–1893)

Dried flowers are considered as stimulant, used in bowel complaints and haemorrhages and is also administered in menorrhagia and seminal weakness. Powder of flower is sprinkled over ulcers and wounds. Paste of flowers is used in the treatment of cough. Flowers also enter into ointment, used on pustules of small pox. Wood is used for tool handles and also as fuel.

(Anonymous, 1948–1976)

Ambasta (1986) recorded similar uses as those recorded by Anonymous (*loc.cit.*).

Roots are used in rheumatism. Bark powder is used on burns and wounds. Flowers are used in the treatment of cancer, cough, cooling, diarrhoea, dysentery, injuries, loss of appetite in pregnancy, menorrhagia, nausea, and also used as tonic. Entire plant is used to treat bone fracture, carbuncle, chancre, dropsy, fever, muscle pain, night blindness, puerperal fever, sores, spleen complaint, syphilis. Leaf paste is applied on sores and ulcers of cattle. Flowers are edible. Leaves are cooked and eaten as vegetable and drinks.

(Jain, 1991)

Critical Notes

All applications noted by the present investigators are new to the science of ethnobotany.

**Wrightia tinctoria* R.Br.

Family: Apocynaceae, Local Name: Kalakuda, Kudai

Ethnobotanical Notes

Medicine

1. About 20–25 ml of infusion of stem bark is administered to mothers for a week to increase lactation.
2. Stem bark is used as antidote against snake-bite.

Veterinary Medicine

1. Fresh branches are tied on horns of cattle as a treatment for vomiting and septics on neck.

Notes from Classics

Root and stem bark have powerful, antidysentric, astringent, febrifuge and anthelmintic properties, used in treatment of fever, dysentery and dropsy. Fruits are used in snake-bite to allay swelling and irritation.

(Watt, 1889–1893)

Bark and seeds are used in flatulence and bilious affection. Dried and ground bark is rubbed over body in dropsy. Decoction of leaves and bark is taken as stomachic. Fresh leaves are chewed to relieve tooth-ache. Flowers are used as vegetable. Tender leaves, pods, and seeds are eaten. Leaves are eaten by cattle, sheep and goats.

(Anonymous, 1948–1976)

Ambasta (1986) recorded uses similar as those recorded by Anonymous (*loc.cit.*) except uses of wood. Wood is used for making cups, plates, combs, pen holders, pencils, carving, frames, small boxes, bobbins and mathematical instruments.

(Ambasta, 1986)

Roots are considered as laxative. Bark is used on bronchitis and piles. Leaves are used in asthma, throat pain and tooth-ache. Plant is used in dysentery. Latex is used on bleeding injuries. Bark and root are used on scorpion sting or snake bite. Bark is used on snake bite to animal. Flowers are eaten as vegetable.

(Jain, 1991)

Critical Notes

Applications of stem bark and branches recorded by the present investigator are not reported in past.

**Xanthium strumarium* Linn.

Family: Compositae, Local Name: Ghokarukata

Ethnobotanical Notes

Medicine

1. Tender leaves are chewed and kept in jaws to treat tooth-ache.

Notes from Classics

Fruits are considered as cooling and used on small-pox. Fruit ash is used as application to sores and lips and mucous membrane of mouth. Prickly involucre is applied to ear or tied in bunch to ear-ring, to cure hemicrania.

(Watt, 1889–1893)

Root is bitter tonic, employed in cancer and scrofula. Root extract is used on ulcers, boils and abscesses. Leaves are reported to possess astringent, alterative, antisyphilitic and diuretic properties. They are also given for scrofula and herpes. Seeds are used on inflammatory swelling and seed oil is used for bladder affections, herpes and erysipelas.

(Anonymous, 1948–1976)

Root extract is applied to ulcers, boils and abscesses. Roots are bitter tonic, used in the treatment of cancer and scrofula. Leaves are astringent, diuretic, and anti-syphilitic, used in scrofula and herpes. Buds are tonic, diuretic and sedative, used for resolving inflammatory swellings and their oil for bladder affections, herpes and erysipelas. Fruits are cooling and demulcent. Herb is diaphoretic, diuretic, emollient and sedative. Decoction of herb is used in chronic malaria, leuccorrhea and urinary diseases. Leaves are used as pot-herb. Seed oil is edible, also used in various soap makings and paints etc. Seed cake is used as fertilizer, toxic if fed to cattle.

(Ambasta, 1986)

Roots are used on boils and ulcers. Leaves are used on malaria, herpes, against ring worms and scrofula. Flowers are useful for tooth-ache. Fruits are cooling and sudorific. Seeds are used to treat cancer and eye diseases. Seed oil is used in rheumatism. Seeds are burnt and fumes are used on piles. Fruits and leaves are edible drinks. Seedlings are eaten as vegetable. Prickles of fruits are removed and garland is put around neck to cure night blindness.

(Jain, 1991)

Critical Notes

Employment of tender leaves for curing tooth-ache is a new report. Plant is exotic, native of Europe.

**Ziziphus mauritiana* Lam.

Family: Rhamnaceae, Local Name: Bor

Ethnobotanical Notes

Medicine

1. Fresh bark is chewed for 4–5 days twice a day to get relief from cough.

Miscellaneous

1. Leaves are fed to goats.
2. Ripe fruits are consumed by tribals.

Notes from Classics

Decoction of roots is used in fever and delirium. Root powder is applied on ulcers and wounds. Leaf paste is applied in strangury and other diseases. Seeds are astringent and used in diarrhoea. Leaves are pounded with leaves of *Ficus glomerata* and applied on scorpion sting. Leaves are used as cattle fodder. Wood is used for making agricultural implements and various small articles.

(Watt, 1889–1893)

Root decoction is used in delirium and fever. Root bark is astringent and used in preparing gargles. Bark is said to be a remedy for diarrhoea. Paste of tender leaves and twigs are used as an application to boils, abscesses and carbuncles. They promote suppurations. Leaves are applied in strangury and other diseases. Fruits are cooling, aperient, astringent, aphrodisiac and fever, haemorrhages, consumption and thirst. Leaves pounded with those of *Ficus glomerata* are applied to scorpion stings.

(Anonymous, 1948–1976)

Fruits of wild plants are cooling anodyne and tonic, employed as an antidote to aconite-poisoning and recommended in nausea and vomiting, also used for abdominal pain in pregnancy. Kernels are sedative, used in insomia. Heart wood is used for turnnery and combs.

(Ambasta, 1986)

Roots are used on wounds and ulcers. Stem bark is used on diarrhoea. Leaves are used to treat eye diseases, conjunctivitis, falling hair, fever and whooping cough. Fruits are used in blood purification, digestion, and stomach-ache. Plant is used in cholera, colic, dysentery, sores, spleen complaints, head-ache and oedema.

(Jain, 1991)

Critical Notes

Bark when fresh is useful to treat cough, an application hitherto unreported.

**Ziziphus nummularia* (Burm.*f.*) Wight & Arn.

Family: Rhamnaceae, Local Name: Kathbor

Ethnobotanical Notes

Miscellaneous

1. Leaves are used as goat fodder.
2. Ripe fruits are edible.

Notes from Classics

Fruits are cooling and astringent, useful in bilious affections. Bark poultice is used for foul sores. Decoction of bark is used as a gargle in sore-throat and ulcerated gums. Fruits are eaten by poorer classes in times of food scarcity. Leaves are valuable fodder for camels, goats, buffaloes and cows. Branches are used for fences.

(Watt, 1889–1893)

Leaves are used in scabies and other skin diseases. Smoke of dried leaves is inhaled as a treatment for cold and cough. Fruits are cooling and astringent, useful in bilious affection. Leaves fallen are dried and used to feed cattle, camels, sheep and goats. Branches are used for fencing.

(Anonymous, 1948–1976)

Dried leaves are smoked on cough and cold. Leaves are used to apply on cutanous diseases. Fruits are cooling, astringent and used in bilious affections. Leaves are used as fodder for cattle. Wood is used as fuel and it is good for charcoal. Seeds are used as bait after saturating with poison to kill gerbils.

(Ambasta, 1986)

Roots are used on diarrhoea. Leaves are applied on boils. Fruits are used on cold and indigestion. Stem bark is used on inflammation of gum. Roots are used as veterinary medicine and for calf-laying.

(Jain, 1991)

**Ziziphus rugosa* Lam.

Family: Rhamnaceae, Local Name: Toran

Ethnobotanical Notes

Miscellaneous

1. Ripe fruits are edible.

2. Leaves are fed to goats.

Notes from Classics

Bark powder mixed with ghee is applied to swellings in cheek caused by tooth-ache and also given in ulcers of mouth. Bark is astringent and used in diarrhoea. Wood is used as fuel, branches are used for fences. Flowers of same plant and petioles of betel and lime are used for menorrhagia. Fruits are eaten by natives in times of scarcity.

(Watt, 1889–1893)

Bark is used to treat diarrhoea. Bark powder mixed with ghee is applied on check swellings caused by infection of teeth. Flowers are used for menorrhagia.

(Anonymous, 1948–1976)

Uses recorded by Ambasta (1986) are similar as given by Anonymous (*loc.cit.*).

Roots are used on dysentery. Stem bark is used in diarrhoea. Flowers are used in mennorrhagia. Entire plant is used in carbuncle, against gum trouble, syphilis and tongue ulcers. Fruits are edible.

(Jain, 1991)

Critical Notes

Fodder value of its foliage is unnoted yet.

Chapter 3

Origins of Vernacular Plant Names

(A) Names after Animals

1) *Abelmoschus esculentus* (L.) Monench. (Malvaceae): 'Waghacha Panja' (Waghacha-Tiger's; panja-palm). Leaf lamina resembles foot palm of a tiger.
2) *Arisaema tortuosum* (Wall.) Schott. (Araceae): 'Saradah-Jad' (Sarada-lizard; jad-plant). Leaf sheath is often mottled with purple, a feature also noted in case of lizards.
3) *Erythrina variegata* L. (Papilionaceae): 'Gadha-palas' [Gadha-donkey; *palas–Butea monosperma (Lam.)* Taub.] Grey bark is similar to the colour of a donkey and the trifoliate leaves are similar to those *Butea monosperma* (Lam.) Taub.
4) *Ixora brachiata* Roxb. (Rubiaceae): 'Khura' (Khur-hoop). The didynamous fruits are shaped and notched like the hoofs of cows or buffaloes.

(B) Name after Animal Feed

1) *Aspidopterys cordata* (Heyne ex Wall.) A. Juss. (Malpighiaceae): 'Bokadveli' (Bokad-male goat; vel-climber). This climber is the favourite fodder of goats.

(C) Names after Medicinal Utility

1) *Alstonia scholaris* R.Br. (Apocynaceae): 'Saitan' (Saitan-ghost). One behaves like a ghost when suffering from irritation. In this case, stem bark is administered to cure the patient, hence the name.
2) *Cissus quadrangularis* L. (Vitaceae): 'Had-sandhi' (Had-bone; sandhi-joint healer). Noded stems resemble bone joints. This plant is also used for treating bone fractures.
3) *Clerodendrum serratum* (Linn.) Moon (Verbenaceae): 'Dati' (Dat-tooth). Roots are rubbed on gum of children for better teething.
4) *Nervilia plicata* (Andr.) Schlect. (Orchidaceae): 'Dudri' (Dudh-milk). The bulbs are given to milching animals to increase lactation.
5) *Stemodia viscosa* Roxb. (Scrophulariaceae): 'Ramban' (Ram-Lord Rama; ban-arrow). It is believed that Lord Rama's arrow never missed the target. Entire plant is used in initial stages of leprosy as a panacous medicine.
6) *Ventilago denticulata* Willd. (Rhamnaceae): 'Ardhashish' (Ardh-half; shish- head). Half head-ache *i.e.*, migraine is believed to be cured after wrapping arms with twigs of this plant.

(D) Names after Relationships

1) *Aeschynomene indica* L. (Papilionaceae): 'Jungli Chinch' (Jungli-wild; chinch–*Tamarindus indica* L). Leaves are similar to those of *Tamarindus indica* L..
2) *Boerhavia diffusa* L. (Nyctaginaceae): 'Ran-ratal' (Ran-wild; ratal-roots *Ipomoea tuberosa* L.). The fusiform roots of this species resemble to those of *Ipomoea tuberosa*.
3) *Hygrophila serpyllum* (Nees) T. And. (Acanthaceae): 'Bhuikarvi' [Bhui-earth; karvi- *Carvia callosa (Nees)*

Bremak.]. It resembles somewhat *Carvia callosa with* respect to the leaves and flowers, but being procumbent lies near the earth, hence the name.

4) *Butea superba* Roxb. (Papilionaceae): 'Palasvel, Vadha' [*Palas-Butea monosperma* (Lam.) Taub.; vel-climber, liana. This liana resembles in all respects *B.monosperma*, except habit, which is climbing.

5) *Piper trichostachyon* (Miq.) DC. (Piperaceae): 'Ran Nagali' (Ran-wild; nagali-leaves *of Piper bettle).* The leaves are similar to those of *P. bettle* and the former is thught to be a wild relative of the latter.

6) *Spondias pinnata (L.f.)* Kurz (Anacardiaceae): 'Ambada' (Amba-mango *i.e.*, fruits of *Mangifera indica* L.). Alluding to the similarity of the fruits with mangoes.

7) *Crotalaria pallida* Ait. (Papilionaceae): 'Jungli Tag' (Jungli-wild; *tag-Crotalaria juncea* L.). Habit foliage, flowers and pods all resemble *Crotalaria juncea* and therefore thought to be its wild relative.

(E) Names after Shape

1) *Careya arborea* Roxb. (Lecythidaceae): 'Kumbhi' (Kumbh-a collared earthen water pot). The fruits resemble earthen pots.

2) *Desmodium gangeticum* (L.) DC. (Papilionaceae): 'Asud' (Asud–whiplash). The elongated inflorescences resemble whiplash.

3) *Leea macrophylla* Roxb. ex Horn. (Vitaceae): 'Supad' (Sup-a wide pan to sift grains by shaking). Leaves are large and shaped like a 'sup'.

(F) Name after Appearance

1) *Helicteres isora* L. (Sterculiaceae): 'Ati' (Ati-threads). The follicles are twisted and show spiral threads.

(G) Names after Habitat

1) *Ficus arnottiana* (Miq.) Miq. (Moraceae): 'Khadak Payar' (Khadak-rock; *payar-Ficus virens Ait.)* Payer-like tree flourishes well especially in rocky habitats.

2) *Rotala densiflora* (Roth ex R.&S.) Kochne (Lythraceae): 'Jalmukhi' (Jal-water; mukh-mouth). The plant grows marginally along the water courses or bodies.

(H) Name after Habitat and Appearance

1) *Dendrobium barbatulum* Lindl. (Orchidaceae): 'Zadya Losan' (Zad-tree; losan-bulbs of *Allium sativum L.).* This plant is usually found on trees and the stem being bulbous resemble the bulbs of *Allium sativum.*

(I) Name after Animal Habitat

1) *Asparagus racemosus Willd.* (Liliaceae):'Sashachi-ghongadi' (Sasa-hare; ghongadi-a coarse woolen blanket). This plant spreads on other plants simulating a woolen blanket. It is also a place for hiding of a hare.

(J) Names after Similarity with Other Plants

1) *Cardiospermum halicacabum* L.(Sapindaceae):'Kali Gunj' (Kali-black; gunj -seeds of *Abrus precatorius* L.). Seeds are similar to those of *Abrus precatorius* but differ in total black colour.

2) *Melia composita* Willd. (Meliaceae): 'Nimbara' *(Nim-Azadirachta indica* A.Juss.) Leaves, fruits and general look are similar to *Azadirachta indica.*

(K) Names after Specific Feature

1) *Cissus setosa* Roxb. (Vitaceae):'Khajgolicha-vel' (Khaj-itching; vel-climber). The entire climber is acrid and clothed with bristly hairs. It causes itching on touch or contact with the plant.

2) *Girardinia diversifolia* (Link.) Friig (Urticaceae): 'Agya' (Agya-flamed, pungent). Flowers and fruits are studded with stinging hairs. If touched cause itching.

(L) Name Based on Utility as Food

1) *Caralluma adscendens* (Roxb.) R.Br.var. *fimbriata* (Wall.) Grav. and Mayur (Asclepiadaceae): Vanjari bhaji (Vanjari–A tribe in the state of Maharashtra; bhaji-vegetable). Stems

are used as vegetable. Vanjari ladies fix a long stick like 'chundo' on their heads. The stems of this plant resemble it.

(M) Name after Taste

1) *Cayratia auriculata* (Wall. ex Wight and Arn.) Gamble (Vitaceae): 'Ambat-vel' (Ambat–sour; vel-climber). Alluding to the sour taste of leaves.

(N) Name after Hardness

1) *Ixora pavetta* Andr. (Rubiaceae): 'Lokhandi' (Lokhand-iron). Stem is hard like iron and hence used as a walking stick while trekking.

(O) Name after Colour

1) *Securinega virosa* (Roxb. ex Willd.) Baill. (Euphorbiaceae): 'Pandhar Phali' (Pandhara-white; phal-fruit). Fruits when ripe are white.

(P) Name Suggestive of Phenology

1) *Indigofera tinctoria* L. (Papilionaceae): 'Diwali' (Diwali-A Hindu festival of nocturnal illuminations). Plants bloom during the days of Diwali festival and the flower clusters are also very bright.

(Q) Name after Morphology

1) *Cayratia trifolia* (L.) Domin (Vitaceae): 'Khandivel' (Khandi-noded; vel-climber). The climbers are conspicously noded.

(R) Name after Height

1) *Acacia auriculiformis A. Cunn* ex Benth. (Mimosaceae): 'Akashya' (Akash-sky). The tallness of the plant finds place while coining the name.

(S) Names after Size

1) *Ailanthus excelsa* Roxb. (Simaroubiaceae): 'Maharukh' (Maha-large; rukh-tree). The trees attains large size.

2) *Ipomoea turbinata* Lag. (Convolvulaceae): 'Bhawara' (Bhavara-top, a toy). Pedicel in fruit is swollen and assumes the shape of a top, a toy for spinning.

(T) Name Based on Ethnobelief

1) *Cassine albens* (Retz.) Kosterm. (Celastraceae): 'Bhutyakalaṣ' (Bhut-ghost; kalas- a pinnacle, a highest point). The tribals believe that the very possession of the twig on head of a person or a house wards off the evil spirit.

(U) Name after Sound

1) *Crotalaria filipes* Bth. (Papilionaceae): 'Phat-phati' (Phat-phati–A motor bike which creates sound like 'Phat'). Pods if pressed while walking under foot makes a sound and hence the name.

(V) Name after Conspicuous Feature

1) *Lantana camara* L. (Verbenaceae): 'Gultura' (Gul-flower; tura-inflorescence). The inflorescence is conspicuous due to bright colour, hence the name.

(W) Name after Change in Colour

1) *Gloriosa superba* L. (Liliaceae): 'Panchrang' (Panch-five; rang-colour). Flower colour changes gradually five times from green, to greenish to yellow to yellow red and ultimately to bright red.

(X) Name after Root Length

1) *Hemidesmus indicus* (L.) Schultes (Asclepiadaceae): 'Anantmul' (Anant-unlimited; mul-root). Roots penetrate deep into the soil.

Chapter 4

Discussion

This study contains the results of in-depth ethnobotanical investigations in a botanically fairly rich district of state of Maharashtra. The observations are extended since June 1997 covering 13 Tahsils* of Nasik District (Maharashtra). The area under investigation was combed ethnobotanically carefully in the hilly tribal areas as well as the plains of the district. This contribution included information on plant uses by the local people pertaining to the food, fodder, oil, medicine, timber, fuel wood, fibre, gum, alcohol, and some other miscellaneous uses. An attempt is made to tap information on concrete as well as abstract uses in the region. The geography, geology, soils, climate, vegetation and information of some tribes is particularly included with a view that these natural and social facts have directly or indirectly bearings on the natural man-plant relationships in the said region. The ancient plantlore has passed through generations orally in tribal and non-tribal societies. The anthropogenic pressure, in general, is steadily increasing in our country on account of demand for agricultural land, industrialization and also due to agricultural practices and the population pressure. The forests are being exterminated and tribal cultures are concomitantly changing. These forces will obviously lead us losing for ever the wealth of empirical knowledge of these ethnic people have had and needs strategies for the conservation of both these plants as well as ethnic groups for posteriority.

A perusal of data contained in the book reveals that total of 364 species belonging to 298 genera and 92 families are in use for different purposes. Other 19 species supplement the drugs used. The dicotyledons are represented by 75 families, 258 genera and 313 species and monocotyledons represented by 47 species belonging to 35 genera and 13 families. Four species belonging to four families of pteridophytes are also included. The exotic plant species number about 46 (Appendix-XIII). Total 34 species of crop weeds are found in use for food, medicine etc. out of which eight species are exotic (Appendix XII). About 21 species are being reported useful for the first time (Appendix XIV). As many as 238 species have been recorded with additional uses. The different uses of plants or their products noted in the district are discussed in the following:

It is to be noted that the tribal and rural people particularly still depend on the indigenous system of medicine. The traditional folklore is being transmitted orally from one generation to another. Their knowledge about medicinal utilities of plants have not received due attention in past. They have remained practicals unknown or lesser known to the public.

Indigenous healthcare system has always played vital role in the discovery of novel products from plants as chemotherapeutic agents, for example, quinine from *Cinchona*, morphine from *Papaver somniferum*, cocaine from *Erythroxylum coca*, and picrolive from *Picroriza kurroaa*. Non-availability of satisfactory drugs in modern systems of medicine for the treatment of various diseases have prompted the scientists to study ethnomedicinal claims. Due to many side effects of drugs of modern system and their high costs, the traditional medicines are being reviewed all over the world. This "Green Wave" is also discernible by the fast developing herbal industries. On account of lure of modern civilization, the traditional data on time-tested remedies appear to have depleted and also continue to be so.

The utility of folk medicines or traditional medicines is widely spread particularly in aboriginal people. On account of lack of health centers, doctors and the expenses for their treatments, there is a high percentage of tribal and rural population depending on it. The different plant species are used to treat different human and animal ailments or diseases even in the modern period. It appears necessary to screen them on modern scientific grounds.

It is amply clear from the contents of this contribution that plant wealth of this district contains a great variety of useful plants, which have been very valuable source of folk medicines to the masses. They employ different plant parts like roots, stems, other underground parts, leaves, flowers, fruits, seeds and their products. The methods of administration of medicine are different. They use raw plant parts or products, juice, extract, decoction, infusion, paste, ash, poultice, fumes or vapours, oil, contact drugs to body parts, bath by adding drugs in bath water etc. Sometimes, the administration of medicine is simultaneously followed by muttering hyms. Probably, it may be a psychological treatment given by the medicine-men. Sometimes, instead of sole drug, two or three drugs in combinations are administered. In such cases, it is difficult to decipher the actually more efficacious plant or drug for a particular disease or a ailment. It is only chemical screening and clinical studies that can earmark accurately.

The inhabitant tribals and some rurals associate supernatural powers of spirit, whether good or bad, with life activities like disease and death, birth and marriage, house building and agriculture etc. The consequences of infringement of some taboos, in their opinion can lead to disasters while treating diseases. One is not able to make a sharp distinction between the so called medical treatment and magical cure. For example, (i) putting twigs of *Calotropis procera* on the roof of a house for easy delivery. (ii) keeping a stem piece of *Urena lobata* in pocket or under cap as a safe-guard from wild animals. (iii) use of latex of *Holarrhena pubescens* mixed with rice grains consumed as a preventive measure against snake-bite etc.

In tribal dominated hilly areas the diseases or complaints of stomach are commonly observed. These are diarrhooea, dysentery, flatulence, indigestion, bleeding stools, gastro, intestine torsion, stomach worms and other forms of stomach-ache. However, many-a-times, the tribal people could not make difference in diarrhoea and/or dysentery and use these terms in a loose sense. Diarrhoeal diseases fall usually into three different categories *viz.*, persistent diarrhoea, acute watery diarrhoea, dysentery and bleeding stools. The conditions are serious in tribal hilly areas of Nasik District as very few people receive the modern healthcare facility in time. People have to resort to the indigenous phytotherapeutic method of treatment in this region.

Our observations particularly on these diseases or ailments will help to manage diarrhoeal management practices. These complaints are particularly connected with the quality of drinking water. The goals of different government programmes in this regard are still to be achieved in this areas.

Apart from the above stomach complaints, the present investigators noted many other diseases or ailments like cough, cold, fever, cholera, urino-genital diseases, skin diseases, dental and eye complaints, ear-ache, head-ache, rheumatism, and problems like bone fracture, abortion etc.

Besides the human diseases, the tribal and rural people of district have also considerable knowledge for treating the diseases prevailing for the livestock. The veterinary diseases like dysentery, ephemeral fever, maggot infected sores, skin diseases, yoke galls, insect bite, lack of milk secretion, problems of delivery, bone fracture etc. are being treated by the people of this district using locally available drugs of vegetable origin.

Few exotic plant species like *Adansonia digitata, Agave vera-cruz, Annona squamosa,* etc.(Appendix No.XIII) have gone down into the local uses. Such plants may also add new medicinal avenues as they were not tapped as such in past in their country of origin. Few ethnomedicinal practices are associated with some beliefs, for instance, the day on which the medicine should be given or passing of the drug over the body of animal in a particular direction etc. It is also a common knowledge that inspite of availability of modern medicines, rural people especially the ethnic tribes of the district, continue to take recourse to their own traditional therapy.

India won the battle on USA patent on turmeric powder. This success story is being viewed because of its many ramifications. It is probably the first case where the use of traditional knowledge base has led to complete cancellation of an unreasonable USA patent. The treasure house of useful medicinal plants is available in our surroundings. It has now become necessary to document and systematize the traditional indigenous knowledge which will in future help prevent bio-piracy.

Though large mass of data exist on herbal drugs, the claims are not to be supposed to be very authentic. Ethnomedicinal plants can be, no doubt, help to discover new drugs or lead molecules for the development of new drugs, provided the data are scientifically evaluated. The isolation of active principles and potent anti-microbial activity of these plants should be studied to validate the claims of the traditional medicine-men. The plant species useful as animal medicine are enlisted in the Appendix-VI.

The increasing demand of food with growing human population necessitates identification of new plants as sources of food. The tribals inhabiting the forest area and even the rural people consume a number of wild plants which are unknown and inaccessible to urban elite. Some plants may have high nutritive value. Tribals are both vegetarian and non-vegetarian. They depend on cultivated crops as well as natural forest products for their day-to-day uses. The food stuff produced from cultivated sources is not sufficient and therefore they have to depend on forests particularly for food. They use these sources as alternative food or sometimes during scarcity. It is observed that use of the plant generally depends upon the availability of plants near their villages. The district is inhabited by large number of tribes. They posses good knowledge on wild edible plants. In the present contribution total 110 species have been noted useful by them. The plant parts mostly used by them are underground parts like tubers, bulbs, rhizomes, corms roots, leaves, flowers, fruits and seeds etc. They are enlisted in the Appendix-I. Toddy a nutritious drink is obtained from the trees namely *Phoenix sylvesteris* and *Borassus flabellifer*. It is also sold in market in fermented form as "Nira". The tribal and rural people rear live-stock. They are fed from the agricultural residues, however, this source does not suffice their total need. Therefore, they have to derive the fodder from crop-weeds or wild forest plants. They are about 92 species useful as fodder presently noted. They are enumerated in the Appendix-IV.

The rural people in the district usually raise flat roofed 'Dhabha" houses. Mostly teak wood (*Tectona grandis* L.) is employed for their construction. The tribal and poor rural people erect huts with sloping roofs. They employ timber obtained from various wild tree species and also the stem and leaves for thatching purposes. The timber yielding 21 species are used for house construction in Appendix II(A) and 16 species are used for thatching huts are enlisted in Appendix-II(B).

Traditionally the people in the district use tender stem or branch twigs as tooth brushes. The different species have peculiar taste and flavour. They may possess medicinal, cleaning and disinfecting properties. This method of cleaning teeth also increases the strength of gums. The commonly used species for these purpose are *Acacia nilotica* (Linn.) Delile *Azadirachta indica* A.Juss. *Baliospermum montanum* (Willd.) Muell-Arg. *Ficus racemosa* L. and *Jatropha curcas* L.

There are some tree species naturally occurring in the district. *Acacia nilotica* (L.) ssp. *indica* (Bth.) Brenan, *Albizia procera* (Roxb.) Bth., *Albizia lebbeck, Anogeissus latifolia* (Roxb. ex DC.) Wall. ex Guill. & Perr. *Boswellia serrata* Roxb. ex Colebr., *Butea monosperma*, (Lam.) Taub., *Lannea coromandelica* (Hout.) Merr., *Sterculia urens* Roxb., *Terminalia crenulata* Roth., *Pterocorpus marsupium* Roxb. var. *acuminatus* Prain. Some of these are important as edible or medicinal. Majority of them are used for glueing purpose.

The fruits of *Sapindus laurifolius* Vahl are employed by the local people for hair washing. Very few trees have been noted by the present investigators in the forest areas or district. This is probably so because nearly all fruits are generally send to the markets. These naturally reduces regeneration of this species. It can be recommended for afforestation programme or captive plantation in the region to better the economy of poor tribal people. *Balanites aegyptiaca* (Linn.) Delile is another species yielding fruits used for washing clothes by the tribals.

Madhuca longifolia (Koen.) MacBride var.*latifolia* (Roxb.) Chevalier is a key-stone species from socio-economic point of view. It provides food and medicine for local tribal people, apart from its use for distillation of wine. These trees are usually found near human habitations. They are mostly planted by tribal people. All the fruits and seeds are collected for extraction of oil. This has hindered natural seed dispersal which resulted in a negative impact on wild population of this economically important species. They are useful only for the tribal people. Possessions, consumption and sale of wine distilled from its flowers are all crime according to Bombay Prohibition Act, 1959. It is, therefore, not usually planted by the non-tribal people in the district. The tree is although of economic value of many counts, no care has be taken to propagate it and is always excluded from the afforestation programmes even in the deforested tribal dominated areas. The present authors feel that it should be given place in social forestry programmes especially in this areas to boost economy of tribals.

Collections of tendu leaves (*Diospyros melanoxylon* Roxb.) fetch good returns for local tribals. The leaves of palas [*Butea monosperma* (Lam.) Taub.] and of Karwal (*Dillenia pentagyna* Roxb.) are usually collected to prepare dinning platters and bowls. The number of latter species are very few in the region. It is, therefore, to be recommended for captive plantation especially near hilly tribal areas. This will give large employment for the tribals.

To supplement the food requirement the tribals usually trap fishes from rivers, rivulets and other water bodies. They use roots, barks, leaves and fruits of different easily and locally available plant species. These are mentioned in the Appendix-V.

Apart from the above uses some plants are employed by the inhabitants of Nasik district for different purposes such as–(i) Hedge (Appendix-III), snake-bite and scorpion sting (Appendix-VII), agricultural implements (Appendix-X), firewood (Appendix-XI), Ethnobelief (Appendix-VIII) and religious (Appendix-IX).

The present authors observed some species deserving urgent conservation. These species are over-exploited by the local people for various purposes. Their indiscriminate collections appear to have resulted in rarification of these taxa. Some of them are follows: The flowers of *Curcuma pusedomontana* Grah. are collected on large scale and marketed at Trimbakeshwar. They are offered to Lord Shiva's idol by the devotees. Secondly, rhizomes are used to increase lactation and to better digestion in women. As the inflorescences are collected before fruit-set, seeds are not dispersed in its natural habitats. Removal of rhizomes is another practice to destroy the entire plants. These obviously check its regeneration. It needs public awareness. The local people should bring this species under cultivation and exploit as a commercial crop. It will fetch a handsome returns to the tribals if cultivated.

The showy attractive flowers of *Gloriosa superba* L. are also collected on large scale and marketed at Trimbakeshwar as an offering to Lord Shiva's idol. Its tubers are used to increase lactation in cattle and women. The flowers are easily collected in months of August and September. This results in massive destruction of flowers, which thereby reduces formation of fruits and seeds. This appears to have culminated into shrinkage of populations of the species.

The male flowers and the tender green fruits of *Dioscorea wallichii* Hook. *f.* are used for vegetable. Collection of male flowers naturally checks pollination and fertilization. Collection of green young fruits minimizes fruit or seed dispersal. This obviously appear to have depleted the species.

The bulbs of *Pancratium sanctae-mariae* Blatt. & Hallb. are collected by the tribal rice cultivators. These bulbs are cut and placed in rice fields. In their opinion, these keep the crop away from diseases, bad omens and help to increase its crop yield. These farmers now-a-day complain for less availability of these bulbs. These appear to have been over exploited. This is an ethnobelief. Public awareness especially in tribal rice cultivators is an urgent need of a day. Likewise, the tubers of *Pueraria tuberosa* (Roxb. ex Willd.) DC. are collected to make bread while taking out tubers. The entire plants are uprooted and destroyed. It is, therefore, few plants are observed in forest areas. It has become necessary to check its further over-exploitation.

Forest are being denuded on account of fires illicit cuttings, over grazing, shifting cultivation, fuel wood collections and the practice of, "Rabani" prevalent in these district. During the practice of "Rabani" branches of some tree species are stripped off. They are burnt in ploughed fields before rainy season. These species are usually *Syzygium cumunii* (L.) Skeels, *Terminalia crenulata* Roth, *Carissa congesta* Wight etc. Thus rapid depletion of forests has obviously become a serious concerned in these region too.

The present authors also enquired especially the origins of vernacular plant names. Some of them appeared very interesting. About 44 vernacular names of plants have been analyzed etymologically and are categorized into different 24 categories. They proved to be good source of ethnobotanical knowledge of the district. These names are good indicators of past notions, concepts and ancient understanding about the ambient vegetation. They are usually founded on plant features such as colour, shape, size, height, taste of plants or their parts. They are also similarly based on features like sound, appearance, similarity with other plants, belief, relationship, phenology, habitat, morphology and even animal features. This type of study obviously shows indepth knowledge about plants of the tribal and rural people of the district.

Chapter 5
Summary

The study on ethnobotanical aspects was initiated since June 1997 and the field work was continued upto July 2002. An effort has been made to collect relevant data on the traditional or folk medicines as well as on the other uses such as food, fodder, fibre, oils, timber, gums etc. employed by the different tribes and rural people of the Nasik district. The plants associated with the religious faiths and beliefs are also noted during the field works. Continued personal attachment with the tribal people has made possible to gather sufficient information on their culture, life style, superstition etc. and more particularly application of plants or plant products for curing human as well as animal diseases and ailments in the region.

The first-hand information is gathered mainly for 364 species belonging to 298 genera and 92 families of plants. Out of these, four taxa belong to pteridophytes. Other 19 species are used as supplementary. As many as 34 species are crop weeds. Totally 46 taxa are exotic in nature. The rich vegetation the district provided the habitants enormous resources of plant-wealth. There was no written record of these resources. Information on the utilization of plant for various purposes passed on through oral communication only. These have been critically tapped from them. Many plant species are useful multipurposely. About 225 species are used as ethnomedicines. The other uses noted are for edible (110 species), house and hut construction (37 species), hedge (10 species), fodder (92 species), fish poison (12 species), veterinary medicines (61 species), antidote for snake bites (20 species), scorpion sting (06 species), ethnobeliefs (20 species) and religious (11 species), agricultural and other implements (16 species), and Firewood (17 species). Vernacular plant names of 44 species have been studied etymologically in ethnobotanical perspective.

It has been possible to make inventory of 21 species of plants as new claims in ethnobotany. There is no earlier report of their utilization in classic literature consulted (Appendix-XIV). Besides these, there are as many as 238 species of plants which have been recorded with additional uses or new applications hitherto unrecorded in classical literature. Screening of these plants for their chemical composition and other potentialities are essential to identify the active principles or utilities. These may lead to discovery of new drugs and effective use.

This study will suffice the need for finding out suitable alternate source of medicinal plants. The plants which have similar therapeutic properties have same active principles and may be used as alternate sources.

The accomplishment of tribal and rural people in understanding plant and their properties is result of long and intimate association with their flora and their dependence on them. Since their knowledge is based on experimentation on human beings, although empirical, it warrants careful thinking. It behoves us to take advantages of their in-depth knowledge, which still exist in this region for scientific scrutiny and adoption for posterity. The scientific scrutiny, of course, includes (i) chemical screening, (ii) pharmacological analysis, (ii) biological activity and (iv) clinical trials.

The book also includes information of geography, geology, soils, vegetation, climate, history of the district, tribals in the district, plant based industries, crops etc. These help elucidate the need and prevalent utilities in the region under investigation.

The present study also made some observations on depletion of forest and plant species in particular. With some exceptions, densely forested area are found degraded to a great extent and now transformed into scrub jungles. Illicit felling of trees, encroachment of forest lands, overgrazing, fires, over-exploitation etc. appear directly responsible for cause of depletion of natural habitats. As a result, natural regeneration have become impossible for some species as discussed earlier. These should be prohibited and stringent action should be taken for their collection by the local people. Protection of natural habitats, preservation and even *ex-situ* conservation of these taxa be taken up immediately for perpetual existence.

This study throws light on the popularity of indigenous system of medicine among tribal and rural people and might draw attention of botanists, phytochemists, pharmacologists, and foresters for further scientific study and conservation of some economically important plant species in the region. This type of ethnobotanical field works pave way for building of eco-friendly society where people live in harmony with nature.

Bibliography

Almeida M.R. and S.M. Almeida (1987). Folklore and utility of vascular plants from Savantwadi. In: Proc. All India Symposium on Biology and Utility of Wild Plants, 23–39. Prof. G.L.Shah Commemoration Vol. Department of Biosciences South Gujarat University, Surat.

Ambasta S.P. (1986). The useful plants of India. Publication and Information Directorate, CSIR, New Delhi, India.

Anand Kumar, Kulkarni B.G. and N.P.Singh (1981). Medicinal and economic plants growing at Pune. Maharashtra Vidnyan Mandir Patrica **15:** 64–74.

Anonymous (1948–1976). The Wealth of India. A Dictionary of Indian Raw Materials and Industrial Products. Vol.1–11. Publication and Information Directorate, CSIR. New Delhi, India.

Anonymous (1982). List of medicinal plants deposited in various herbaria of the Botanical Survey of India. *Bull.Bot.Surv. India.* **2:** 180–273.

Anonymous (2001a). Brief information of I.T. D.P.Nasik and Kalwan, District Nasik. Adivasi Vikas Karyalay, Nasik (Maharashtra), India.

Anonymous (2001). Population: District Statistical Department, Nasik (Maharashtra), India.

Ansari M.Y. and M.P.Nayar (1978). Some less–known plants from western ghats useful for essential oil industry. *Indian Perfumer*. **22(3)**: 210–213.

Ansari M.Y. and Rolla S. Rao (1973). *Iphigenia stellata* Blatter (Liliaceae)–Its identity and economic importance. *Bull.Bot. Surv. India*. **15(1–2)**: 118–122.

Badhe P.D. and B.N. Sharma (1981). Ethnomedicinal methods of Korku tribes of west Melghat. *B.M.E.B.R.* **3(2–4)**: 162–172.

Badhe P.D. and V.K.Pandey (1988). A study of medicinal and economic plants of Amaravati Division, Amaravati Circle, Maharashtra. *B.M.E.B.R.* **11(1–4)**: 1–39.

Badhe P.D., Sharma L.K. and Y.K.Pande (1997): Rare folkmedicines practiced by Gond tribes of Wadsa forest division. Gadchiroli district (Maharashtra State). *Bull. Medico–ethno–bot: Res.* **18(1–2)**: 12–17.

Bhamare P.B. (1995). Some anti–venom medicinal plants from tribals of Dhule district Maharashtra. *JAST*. **1(1)**: 36–37.

Bhamare P.B. (1998a). Traditional knowledge of plants for skin ailments of Dhule and Nandurbar district. Maharashtra (India). *J.Phytol. Res.* **11(2)**: 195–196.

Bhamare P.B. (1998b). Ethnomedicinal plants used by tribals in Dhule district Maharashtra, India. *J.Swamy Bot. Cl.* **15**: 18–83.

Bhogaonkar P.Y. and V.D.Devarkar (2001a). Studies in ethnobotany of Korkus of Melghat Amaravati District, Maharashtra. I–Antisterility Drugs. *JAST* 4(1–2)63–66.

Bhogaonkar P.Y. and V.D.Devarkar (2001b). Studies in Ethnobotany of Korkus of Melghat Amaravati District, Maharashtra. II–Drugs for Urino–Genital Disorders. *JAST* 4(1–2) 67–68.

Bhuktar A.S. (2002). Traditional use of medicinal plants of Hingoli district In: Proc. Plant Resource Development. 146–152 (Ed.A.M.Mungikar) Publ. A.M.Mungikar. Dr.B.A.M. University, Aurangabad.

Caius J.F. and K.S. Radha (1939). The gum arabic of bazaars and shops of Bombay. *J.Bombay Nat.Hist. Soc.* **12**: 261–271.

Chandras G.S. (1966). Economic forest products of Maharashtra. In: Proc.The Autumn School in Botany Mahabaleshwar (Ed.T.S.Mahable): 26–36. Department of University of Poona, India.

Chute G.S. and V.J.Tiwari (1999). Indigenous ethnomedicinal plants used by tribal people of Bhandara and Gadchiroli district of Maharashtra State. *Indian J. Nat. Prod.* **15(1)**: 3–8.

Cooke T. (1958). The flora of the Presidency of Bombay. Vol. 1–3. B.S.I. Calcutta (Reprint edn.).

Das P.K. and M.K.Misra (1988). Some medicinal plants among Kondhas around Chandrapur (Koraput). *J.Econ.Tax.Bot.* **12(1)**: 103–109.

Datar Rekha and V.D.Vartak (1975). Enumeration of wild edible plants from Karanala bird sanctuary. Maharashtra State. *Biovigyanam* **1**: 123–129.

Deokule S.S. and D.K.Magdum (1992). Enumeration of medicinal plants from Baramati area, District Pune, Maharashtra State. *J.Econ. Tax.Bot. Addl. Ser.* **10**: 289–299.

Deshmukh V.R. and M.A.Dhore (1994). Nutritive values of some lesser known herbiv food from Melghat tiger reserve Maharashtra. Repr. *The Indian Forester*. **120(10)**: 920–923.

Deshmukh V.R., Muratkar G.D. and S.P.Rothe (1999). Preliminary observation of the medicinal and economically important leguminous plant species from Amaravati Tehsil. *J.Econ. Tax.Bot.* **23(2)**: 283–289.

Ethoven R.E. (1930). Tree and animal worship in western India. *Proc. Jubilee Congress–1926*. London: 181–187.

Gadgil Madhav and V.D.Vartak (1973). Groves dedicated to the gods. *The Illustrated Weekly of India*. 20–21.

Gadgil Madhav and V.D.Vartak (1974). Sacred groves of India–A plea for continued conservation. *J.Bombay Nat.Hist.Soc.* **72(2)**: 314–320.

Gadgil Madhav and V.D.Vartak (1976). The sacred groves of Western ghats of India. *Econ.Bot.* **30**: 152–160.

Gadgil M. and V.D.Vartak (1981). Sacred Groves of Maharashtra: An inventory. In: Glimpses of Indian Ethnobotany (Repr. By S.K.Jain) 279–294.

Gaikwad J.S. (1988). Role of minor forest produce in tribal economy. In: A Review of Tribal Research Studies, Silver Jubilee Volume 103–108 (Ed.Dr.Govind Gare) Tribal Research and Training Institute Pune (M.S.) India.

Ghate V., Agate V. and V.D.Vartak (1988). Promising economic potential of Shumul (*Bombax cieba* L.) as tuber crop. *Indian Journal of Forestry* **11(2)**: 158–159.

Ghate V.S., Vartak V.D. and M.S. Kumbhojkar (1990). Indigenous fuel–wood species from western ghats of Maharashtra. Repr. *The Indian Forester* **116(3)**: 233–237.

Ghate Vinaya S. (1998). Plants in patra–pooja: Notes on their identity and utilization. *Ethnobotany* **10**: 6–15.

Ghate Vinaya S. and Hemlata D. Sane (1994). Notes on socio–economic status of the flora of Kanakeshwar hills, Raigad District, Maharashtra. *Indian J. of Forestry*. **17(2)**: 119–128.

Ghate Vinaya S. and V.D.Vartak (1996). Socio–economic and domestic utilization of forest tree resources by tribal communities in western Maharashtra. In: Ethnobiology in Human Welfare (Ed.S.K.Jain): 329–330. Deep Publications, New Delhi, India.

Giri R.B. (1994). Melghatatil aushadopyogi vanaspati. Technical Bulletin No.5 Directorate, Project Tiger, Melghat Paratwada, District Amaravati, Maharashtra.

Godbole Archana (1994). Less known uses of three climbers from western Maharashtra. *Indian J. of Forestry*. **17(4)**: 351–352.

Godbole Archana (1996). Role of tribals in preservation of sacred forests. IN: Ethnobiology in Human Welfare (Ed.S.K.Jain): 345–348. Deep Publications New Delhi, India.

Godbole S.R. (1934). Some forest trees and shrubs from the Bombay Presidency used as fodder for animals. *Poona Agric. Coll. Mag.* **26**: 19–25, 70–77.

Gunjatkar Nalini and V.D.Vartak (1982). Enumeration of wild edible legumes from Pune district, Maharashtra State. *J.Econ.Tax.Bot.* **3**: 1–9.

Gunjatkar Nalini and V.D.Vartak (1983). Ethnobotanical and floristic studies on Bherli–mad (*Caryota urens* Linn.) from Western India. *J.Econ.Tax. Bot.* **4(2)**: 323–329.

Gupta, V.C., Hussain S.J. and S.Imam (1997). Important folk medicinal plants and traditional knowledge of tribals of Aurangabad and Nasik Forest divisions of Maharashtra (India). *Hamdard Medicus* **40(2)**: 59–61.

Harshberger J.W. (*1885). Some new ideas Philadelphia Evening Telegraph.

Jain S.K. (1987). A Manual of Ethnobotany, Scientific Publishers Jodhpur, India.

Jain S.K.(1989). Methods and Approaches in Ethnobotany. *Soc.Ethnobotanist*. Lucknow, India.

Jain S.K. (1991). Dictionary of Indian Folk Medicine and Ethnobotany. Deep Publication, New Dehli, India.

Jain S.K. and Rao (1977). Hand Book of Field and Herbarium Methods.

Janardhanan K.P. (1961). Fodder grass and legumes of Khed taluka, Poona District. In: Proc.Second Annual Sess. Acad.Agri. Sci. Coimbtore 106–113.

Janardhanan K.P. (1963). An enumeration of the medicinal plants of Khed taluka, Maharashtra State. *Bull.Bot.Surv. India*. **5(3–4)**: 363–374.

Jayananda Tosh (1981). Observations on wild edible plants from hilly regions of Maharashtra State from Hilly regions and future prospects. 261–271. In: Glimpses of Indian Ethnobotany (Ed.S.K. Jain): New Delhi.

Jaynanda Tosh (1996). Ethnobotanical study of western Maharashtra. J.Econ.Tax.Bot.Addl. Ser. 12, Scientific Publishers Jodhpur (India) 169–174.

Jayananda Tosh, Vartak V.D. and M.S.Kumbhojkar (1959). Some wild edible plants from the hilly regions of Pune District. *J.Bombay Nat.Hist.Soc.* **65(1)**: 9–26.

Kamble S.Y. and S.G. Pradhan (1980). Ethnobotany of the 'Korkus' in Maharashtra. *Bull.Bot.Surv.India* **22(1–4)**: 201–202.

Karnik C.R. (1966). Some medicinal plants from Satpura mountains. *Indian Forester* **92**: 173–183.

Khandekar V.P. and G.N.Balve (1998). Local medicinal plants from Koyna valley. *Ayurved update* **1(2)**: 7–9.

Kirtikar K.R. (1906). A note on an edible puff–balls from the Thana district. *Journ.Bombay Nat.Hist.Soc.* **7**: 816–817.

Korekar S.L. and P.B.Papadiwal (2002). Studies on forest based cottage industries in Osmanabad and Latur districts. In: Proc. Plant Resource Development: 142–144 (Ed.A.M.Mungikar and A.S.Bhuktar) Dr.B.A.M. University, Aurangabad.

Kothari M.J. and S. Moorthy (1987). Floristic composition and utility of wild plants in Raigad district, Maharashtra. In: Proc.All India Symposium on Biology and utility of Wild Plants 100–107. Prof.G.L.Shah Commemoration Vol. Department of Biosciences, South Gujarat University, Surat.

Kshirsagar R.D. and Singh N.P. (2000b). Less known ethnomedicinal uses of plants in Jenukuruba tribe Mysore District, Southern Indian *Ethnobotany* **12**: 118–122.

Kshirsagar S.R. and D.A.Patil (2002) Ethnobotanical and Ethnomedicinal observations in Jalgaon District of Maharashtra. *Vanyajati* **3**: 9–14.

Kulkarni (1968). Medicinal plants from Vengurla, Sawantwadi, Phonda and Ratnagiri district. Maharashtra. *M.V.M. Patrika* **3(1)**: 18–24.

Kulkarni D.K. and M.S. Kumbhojkar (1992a). Ethnobotanical studies on Mahadeokoli tribe in Western Maharashtra–Part–I. Cordage Plants. *J.Econ. Tax.Bot.* Addl. Ser.**10**: 111–115.

Kulkarni D.K. and M.S. Kumbhojkar (1992b). Ethnobotanical studies on Mahadeokoli tribe in western Maharashtra–Part–III. Non conventional wild edible fruits. *J.Econ.Tax.Bot.*Addl.Ser.**10**: 151–158.

Kulkarni D.K. and M.S. Kumbhojkar (1992c). Ethnobotanical studies on Mahadeokoli tribe in Western Maharashtra–Part–II. Fodder plants. *J.Econ.Tax.Bot.* Addl.Ser.**10**: 123–128.

Kulkarni D.K. and M.S.Kumbhojkar (1993). Kitchen garden plants of Mahadeokoli tribe is Maharashtra. *Ethnobotany* **5**: 119–127.

Kulkarni D.K. and M.S.Kumhojkar (1996). Pest control in tribal areas of Western Maharashtra on ethnobotanical approach. *Ethnobotany* **8**: 56–59.

Kulkarni D.K. and M.S.Kumbhojkar (1999). Dam-disaster to biodiversity in Sacred grooves. The Deccan Geographer. **37**(1): 65–71.

Kulkarni D.K. and M.S. Kumbhojkar. (2000). Plant conservation in Maharashtra during 17th Century. *Ethnobotany*, **12**: 29–38.

Kulkarni D.K., Agati V.V. and M.S. Kumbhojkar (2003). Leafy Vegetables consumed by Mahadeo Koli Tribe in Western Maharashtra with their Nutritional Potential. *Ethnobotany* **15(1–2)**: 34–39.

Kulkarni D.K., Kulkarni M.B. and M.S.Kumhojkar (1996). Quantification of three reosources utilized for medicinal purpose by Mahadeokoli of western Maharashtra India. In: Ethnobiology in Human Welfare (Ed.S.K.Jain): 429–432. Deep Publications, New Delhi, India.

Kulkarni D.K., Kumbhojkar M.S. and A.S. Upadhye (1995). Ethnobiological resources used in traditional medicines by Mahadeokoli tribe from Western Maharashtra. Flora and Fauna. **1(1)**: 65–66, Surya Publishers, Pune, India.

Kulkarni D.K., Kumbhojkar M.S. and D.S.Nipunge (1990). Research notes (IV) Notes on fish stupefying plants from Western Maharashtra. (Rep.) *The Indian Forester*. **116(4)**: 331–333.

Kulkarni D.K., Kumbhojkar M.S. and Rajeev Khedkar (1998). Rice germplasm collection, conservation and use: A case study in Western Maharashtra. *Ethnobotany* **10**: 27–31.

Kulkarni D.K., Kumbhojkar M.S., Agate V.V., Joshi N.S. and V.N.Joshi (1991). Nutritional contents in *Flocourtia* from Western Maharashtra. *J.Fd.Sci. Technol.*, **28(2)**: 118–119.

Kumar Anand, Kulkarni B.G. and N.P.Singh. (1980). Medicinal and economic plants growing at Pune. *M.V.M. Patrika* **15(2)**: 64–74.

Kumbhojkar M.S. and D.K.Kulkarni (1998). Environmental impact of sacred groves in Western Ghats of Maharashtra. *Science and Culture*. **64(9–10)**: 205–207.

Kumbhojkar M.S. and V.D.Vartak (1988). Ethnobotanicals studies on wild edible grapes from sacred groves in western Maharashtra. J.Econ.Tax.Bot. **12(2)**: 257–263.

Kumbhojkar M.S., Kulkarni D.K. and A.S. Upadhye (1991). Ethnobotany of *Cissus quadrangularis* L. from India. *Ethnobotany*, 21–25.

Kumbhojkar M.S., Upadhye A.S. and D.K.Kulkarni (1996). Religious forest patches among Mahadeokoli tribal localities: Social cultural and environmental relationship. In: Ethnobiology in Human Welfare (Ed. S.K.Jain): 349–351. Deep Publications, New Delhi, India.

Kurion J.C. and B.V.Bhanu (1980). Ethnomedicine: The Nomadic Vaidus of Maharashtra. *The Eastern Anthropologist*. **33(1)**: 71–78.

Kurion, J.C. and R.T.Tribhuwan (1990). The Traditional Medical Practitioners of Sahyadri. *The Eastern Anthropologist*, **43(3)**: 251–258.

Lakshminarasimhan P. and B.D.Sharma (1990). 'Pawari'–an adviasi musical instrument. (Repr.) *Bull. Bot.Surv. of India* **30**(1–4): 179.

Lakshminarasimhan P. and B.D.Sharma (1991). Flora of Nasik District. Bot. Surv. of Calcutta, India.

*Lisboa J.C. (1887). Notes on Mahabaleshwar and other Indian arrow root yielding plants. *J.Bombay Nat.Hist.Soc.***2**: 140–141.pp.1.

Magdum S.M. and R.M.Gejage (2004). Plants used for Respiratory Disorders by Ethnic Tribes in Chandoli Area of Sangali District. In: Proc. National Conference Human Health and Nutrition A Biotechnological. 128–130. Dept. of Botany, B.N.Bandodkar College of Science, Thane (M.S.).

Mahajan S.D. and V.V.Divan (1968). Contribution to the flora of Kolhapur district. Vegetation and forest resources of Radhanagari area. Journal of Shivaji University: 169–75.

Majgaonkar C. and S.G. Phadke (1999). Local medicinal plants from Koyna valley–4. Bhumyamlaki Phyllanthus niruri Linn. *Ayurved update* **2(1)**: 8–9.

Malhotra S.K. and S.Moorthy (1973). Some useful and medicinal plants of Chandrapur district, Maharashtra State. *Bull.Bot. Surv. India.* **15(1&2)**: 13–21.

Mirashi M.V. and J.B.Salpekar (1975). Ornamental plants at Amaravati–I. *Botanique*, **6(4)** 195–204.

Mokat D.N. and S.S.Deokule (2004). Plants used as Veterinary Medicines in Ratnagiri District of Maharashtra. *Ethnobotany* **16(1–2)**: 131–135.

Mudaliar S.K., Kulkarni B.G. and B.D. Sharma (1987). Unpublished ethnobotanical information of G.M.Ryan and G.A.Gammie. *J.Econ.Tax. Bot.* **11(2)**: 311–320.

Naik V.N. (1998). Flora of Marathwada Vol.1–2. Amrut Prakashan, Aurangabad, India.

Naser Rafiuddin and N.P.Vikos (2002). Plants used in ethnoveterinary practices in Aurangabad district, Maharashtra–I. In: Proc.Plant Resource Development. Aurangabad–202–210. (Ed.A.M.Mungikar and A.S. Bhuktar) Dr.B.A.M. University, Aurangabad.

Nilegaonkar Smitha, Vartak V.D. and R.G.Chitre (1985). Nutritional evaluation of some wild food plants from Pune and neighbouring districts, Maharashtra State. Part–I. *J.Econ.Tax.Bot.* **6(3)**: 629–635.

Nipunge D.S., Kulkarni D.K. and V.D.Vartak (1993). Floristic studies on sacred groves from Sinhagad hills in Pune district, Maharashtra State. In: Higher Plants of Indian Subcontinent, (Additional series of Indian Journal of Forestry No.VII). **4**: 153–159. Bishen Singh Mahendrapal Singh, Dehra Dun, U.P. India.

Nipunge D.S., Kumbhojkar M.S. and V.D. Vartak. (1988). Studies on sacred groves of Maharashtra, Part–I. Observation of Sagdara grove in Pune District. *Indian J. of Forestry* **11(4)**: 282–286.

Nipunge D.S., Vartak V.D. and M.S.Kumbhojkar (1992). Cultural heritage of sacred grooves. *Proceedings* South Indian History Congress Pune. 1988. 170–176.

Padhye M.D., Deshmukh V.K. and V.J.Tiwari (1992). Ethnobotanical study of Korku Tribe of Amaravati District, Maharashtra State (India). *Int.Jl.Pharmacogh* **30(1)**: 17–20.

Patil D.A. (1998a). Vernacular plant names: Their origin and utility in Dhule District, Maharashtra. *Ethnobotany*. **10**: 130–132.

Patil D.A. (1998b). Census of genus Ipomoea L. in Dhule District Maharashtra. *J.Phytol.Res.* **11(2)**: 121–126.

Patil D.A. and P.B.Bhamare (1996). Beautiful Forest Trees for Amenity Plantations in Dhule District (Maharashtra). *BRIS JAST* **2**: 58–64.

Patil D.A. (2000) Sanskrit plant names in an Ethnobotanical perspective: I. *Ethnobotany* **7(3–4)**: 229–232.

Patil D.A. (2003) Flora of Dhule and Nandurbar Districts (Maharashtra). Bishen Singh Mahendra Pal Singh, Dehradun, (India).

Patil D.A., M.V.Patil and Shubhangi Pawar (2004a) Ethnobotanical Aspects of Crop Weeds of North Maharashtra. *Ethnobotany* (In Press).

Patil D.A., M.V.Patil and Shubhangi Pawar (2004b) Ethnomedicinal Remedies Against Diabetes In India: A systematic Census. In: Ethnomedicine and Human Welfare Vo. III, pp.90–117. Ukaaz Publications, Hyderabad, India.

Patil D.A., S.C.Borse and P.B. Bhamare (1990) Medicinal plantlore of the tribals of Dhule (Maharashtra). *Biojounal* **2(l): 47–54.**

Patil D.A., Shubhangi Pawar and M.V.Patil (2002). Mahuwa Tree And The Aborigines of North Maharashtra (India) *Natural Product Radiance* **3(5)**: 356–358

Patil M.V. and D.A. Patil (2000a) Some more wild edible plants of Nasik District (Maharashtra). *Ancient Science of Life.* XIX **(3&4)**: 102–104.

Patil M.V and D.A. Patil (2000b) Bases and utility of vernacular plant names in Nasik District (Maharashtra). *J.Swamy Bot.Cl.* **17(3–4)**: 83–85.

Patil M.V. and D.A. Patil (2000c) Folk medicine of Nasik District (Maharashtra), India. *Ancient Science of Life* XXI: 26–30.

Patil M.V. and D.A.Patil (2001) Ethnoveterinary herbal medicines from Nasik District (Maharashtra) *Journal Non–timber Forest Products* **8(1–2)**: *19–24.*

Patil M.V. and D.A. Patil (2002a) Some herbal remedies on stomach complaints in Nasik District (Maharashtra). Proceeding of National Seminar held at Department of Botany, Dr.B.A.Maharashtra University, Aurangabad. (Eds. Prof.A.M.Mungikar and A.S. Bhutkar), pp.123–127.

Patil M.V. and D.A.Patil (2003a) Plants Used In Reproductive Ailments By Tribals of Nasik District. *Ancient Science of Life*. (In Press).

Patil M.V. and D.A.Patil (2003b) Some notable ethnomedicinal plants from Nasik District (Maharashtra). *Indian Journ. Traditional Knowledge* (In Press).

Patil M.V. and D.A.Patil (2003c). Some herbal remedies among the tribals of Nasik District. Natural Product Radiance (CSIR) (In Press).

Patil M.V. and D.A.Patil (2003d) Certain Ethnomedicinal Practices Amongst Aborigines of Nasik District (Maharashtra) For Human Skin Afflictions. In: Advances In Medicinal Plants, Asian Medicinal Plants and Health Care Trust, Jodhpur, India (In Press).

Patil M.V., Pawar Shubhangi and D.A.Patil (2003). Dental and Oral Healthcare In North Maharashtra. *Journal of Non–Timber Forest Products* **11(3)**: 208–210.

Patil M.V., Pawar Shubhangi and D.A.Patil (2004a) Native Herbal Recipes For Respiratory Afflictions From Jalgaon and Nasik Districts of Maharashtra. In: Proc. National Conference on Human Health and Nutrition: A Biotechnological Approach. PP.183–187, Department of Botany, B.N.Bandokar College of Science, Thane (M.S.).

Patil M.V., Pawar Shubhangi and D.A.Patil (2004b) Ethnobotany of *Butea monosperma* in North Maharashtra (India). Natural Product Radiance (CSIR) (In Press).

Patil S.L. and D.A. Patil (2003). Unconventional herbal medicine in Shirpur taluka of Dhule District (Maharashtra), India. Natural Product Radiance (CSIR) (In Press).

Pawar Shubhangi and D.A.Patil (1999). Certain unconventional food plants of Jalgaon District, Maharashtra. *J. of Non–Timber Forest Products* **7(3–4)**: 229–232.

Pawar Shubhangi and D.A.Patil (2000). The origin and utility of vernacular plant names in Khandesh region of Maharashtra. *J.Swamy Bot.Cl.* **17**: 43–47.

Pawar Shubhangi and D.A.Patil (2001) Herbal folk medicines of Jalgaon District, Maharashtra. *Ancient Science of Life*. **21(2)**: 87–91.

Pawar Shubhangi and D.A.Patil (2003) Observations on folkloric medicines of Jalgaon District (Maharashtra). *Indian Journal of Traditional Knowledge*, **3(4)**: 437–441.

Pawar Shubhangi and D.A.Patil (2004a) Little known or Unknown Uses of Naturalised Exotic Plants of Jalgaon District. *J. Non-Timber Forests Products* (In Press).

Pawar Shubhangi and D.A.Patil (2004b). Herbal folk remedies for urinogenital complaints in Jalgaon District of Maharashtra. *Ethnobotany* **16(1–2)**: 86–90.

Pawar Shubhangi and D.A. Patil (2004c) Herbal Folk Remedies Against Rheumatic Disorders in Jalgaon District (Maharashtra) (In Press).

Pawar Shubhangi, Patil M.V. and D.A.Patil (2004). Fish stupefying plants used by tribals of North Maharashtra. *Ethnobotany* **16(1–2)**: 136–138.

Puri G.S. (1959). Survey of medicinal plants in western India. *Indian Journ.Pharm.* **20**: 8.

Puri G.S. and S.K.Jain (1960). Survey of some oil–yielding plants of western India. *Bull.Bot.Surv.India.* **2(1&2)**: 95–98.

Rathor O.S., Zare M.K., Kagne R.M. and V.B. Chavan (2002). Ethnobotanical studies in some less known plants of Marathwada. In: Proc.Plant Resource Development.: 86–89. (Ed.A.M.Mungikar and A.S. Bhuktar) Dr.B.A.M. University, Aurangabad, India.

Rohidas S.B. and G.J.Bankar (2002). Marathwada: Rich source of medicinal plants In: Proc.Plant Resource Development. 224–228. (Ed.A.M.Mungikar) Dr.B.A.M. University, Aurangabad, India.

Rolla Sheshagiri Rao (1964). The doum palms in India. *Botanical Survey of India*, Poona. **8**: 49–54.

Ryan G.M. (1903). Famine foods: Dioscorea pentaphylla an important edible. wild yam of the Thane District, Bombay Presidency, J.Bombay Nat.Hist.Soc. **14**: 772–775.

Santapau H. (1956). The poisonous qualities of *Calotropis gigantea* R.Br. J.Bombay *Nat.Hist.Soc.* **54**: 218.

Sensarma Priyadarsan (1984). Ethnobotanical investigation in the

Indian *Puranas* V.The *Kurma Purna J.Econ.Tax.Bot.* **5(3)**: 634–640.

Shah G.L. (1984). Some economically important plant of Salsette island near Bombay. *J.Econ.Tax.Bot.***5(4)**: 753–566.

Shah G.L., Yadav S.S. and V.Badrinath (1983). Medicinal plants from Dahanu forest divison in Maharashtra State. *J.Econ. Tax.Bot.* **4**: 141–151.

Sharma B.D. and B.G. Kulkarni (1980). Floristic composition and peculiarities of Dev Raies (Sacred groves) in Kolhapur District. *J.Econ.Tax.Bot.* **1**: 11–32.

Sharma B.D. and P.Lakshminarasimhan (1986). Ethnobotanical studies on the tribals of Nasik District (Maharashtra). *J.Econ.Tax.Bot.* **8(2)**: 439–454.

Sharma B.D. and P.Lakshinarsimhan (1996). Ethnobotanical Studies on the Tribals of Nasik District (Maharashtra) *J.Econ.Tax.Bot.* **8(2)**: 439–454.

Sharma B.D. and S.K.Malhotra (1984). A contribution to the ethnobotany of tribal areas in Maharashtra. *J.Econ.Tax. Bot.* **5(3)**: 533–537.

Sharma B.D., Karthikeyan S. and N.P.Singh (1996). Flora of Maharashtra State Monocotyledones. P.Lakshminarasimhan Bot.Surv. of India, Calcutta, India.

Shekhawat G.S. and Sunil Anand (1984). An ethnobotanical profile of Indian desert. *J.Econ.Tax.Bot.* **5(3)**: 591–598.

Shisode S.B. and D.A.Patil (1993). Native medicinal plants of Jalgaon district Maharashtra. *Biojournal* **5(1&2)**: 79–82.

Singh N.P. and S.Karthikeyan (2000a). Flora of Maharashtra State: Dicotyledons Vol.I Bot.Surv.of India, Calcutta, India.

Singh N.P. and S.Karthikeyan (2001). Flora of Maharashtra State: Dicotyledons Vol.II. Bot Surv.of India, Calcutta, India.

Tarafder C.R. (1984). Ethnogynaecology in relation to plants Part–III. Plants used to accelerate delivery and Pre and post–natal complaints. *J.Econ.Tax. Bot.* **5(3)**: 572–576.

Tayade S.K. and D.A. Patil (2003a) Hitherto Untapped Plantlore From Nandurbar District (Maharashtra). *Natural Product Radiance* **4(1)**: 46–50.

Tayade S.K. and D.A. Patil (2003b) Census of Fodder Plants Used by Tribals of Nandurbar District. *Jour. Swamy Bot. Cl.* **21(3–4)**: 81–84.

Tayade S.K. and D.A.Patil (2003c) Ethnomedicinal Wisdom of Tribals of Nandurbar District (Maharashtra). Natural Product Radiance (In Press).

Tekale N.S. (2002). Plant resource development for upliptment of Warli tribe in Thane district of Maharashtra. In: Proc.Plant Resource development Aurangabad. 23–33. (Ed.Am.Mungikar and A.S.Bhuktar) Dr.B.A.M. University, Aurangabad, India.

Tiwari V.J. and M.D. Padhye (1993. Ethnobotanical study of Gond tribe of Chandrapur and Gadchiroli District of Maharashtra State, India. *Fitoterapia*, 64: 58–62.

Tribhuwan, Robin D. and Preeti Peters (1992). Medico–ethnobiology of the Katkaris and Thakurs. *Tribal Res.Bull.* **14**(1): 20–24.

Upadhye A.S. and D.K.Kulkarni (2004). Traditional Phytotherapy for Insect Bites among the tribals in Western India. *Ethnobotany* **16(1–2)**: 59–61.

Upadhye Anuradha, and M.S.Kumbhojkar (1996). Ethnobotany of genus sterculia L. from India *J.Econ.Taxon. Bot*.Add.Ser.12. Scientific Publishers, Jodhpur (India). 40–42.

Upadhye Anuradha and M.S. Kumbhojkar (1992). Ethnobotany of *Madhuca* from western Maharashtra. *J.Econ.Bot.*Add. Ser. **10**: 79–81.

Upadhye A.S. and M.S. Kumbhojkar (1998). *Schleichera oleosa* (Laur.) Oken. Tree. A boon for tribals. *Ethnobotany* **10**: 38–41.

Upadhye A.S., Kumbhojar M.S. and D.K.Kulkarni (1997a). Ethno–medico–botany of some sacred plants of western Maharashtra. *Ethnobotany*. 9: 65–68.

Upadhye A.S., Kumbhojkar M.S. and D.K. Kulkarni (1997b). Medico–botany of genus *Mucuna* from western Maharashtra. *B.M.E. B.R.* **8(1–2)**: 51–54.

Upadhye Anuradha S., Vartak V.D. and M.S.Kumbhojkar (1994). Ethno-medico-botanical studies in western Maharashtra, India. *Ethnobotany* **6**: 25–34.

Upadhye Anuradha, Kumbhojkar M.S. and V.D.Vartak (1986). Observations on wild plants used in folk medicine in the rural areas of Kolhapur District. *Ancient Science of Life* **6(2)**: 119–121.

Upadhye Anuradha, S.Kumbhojkar M.S. and V.D.Vartak (1988). Cultivation of medicinal plants along sea–coast in Maharashtra. *B.M.E.B.R.* **7(1–2)**: 32–39.

Uphadye A.S., Kumbhojkar M.S. and D.K.Kulkarni (1999). Tribal culture and folk medicines from western Maharashtra. *Science and Culture.* **65(9–10)**: 288–289.

Vartak V.D. (1957). Aaushadhi Vanaspati (Katraj ghat and Parvati). (Repr.) *The Ayurvidya* **20(9)**: 1–10.

Vartak V.D. (1959a). Medicinal plants from the hilly regions of Poona and Satara districts, Bombay State, A and B. (*Repr.) Ayurvidya* **23&23**: 11–12.

Vartak V.D. (1959b). Some edible wild plants from the hilly regions of Poona district, Bombay State. *J.Bombay Nat.Hist.* **56**: 8–25.

Vartak V.D. (1959c). Medicinal plants from hilly regions of Pune and Satara districts. Bombay state. *Ayurvidya* **22(11)**: 455–458.

Vartak V.D. (1970). Beautiful wild plants from Dangi hills Maharashtra State. J.Poona Univ. (Sci.& Tech.) **40**: 179–183.

Vartak V.D. (1971). Beautiful wild plants from Dangi hills Maharashtra State. J.Poona Univ. (Sci.& Tech.) **40**: 177–184.

Vartak V.D. (1980). Note on ambemohar pat (*Pandanus amaryllifolius* Roxb.) from Western India. *J.Bombay Nat.Hist. Soc.* **78**: 196–198.

Vartak V.D. (1981a). Observations on wild edible plants from hilly regions of Maharashtra and Goa. 231–241. In: Glimpses of Indian Ethnobotany (Ed.S.K.Jain), Oxford and IBH Publishing Co., New Delhi.

Vartak V.D. (1981b). Relief forest pockets of Panshet water–catchment area Poona district, Maharashtra State. *Biovigyanum.* **7**: 145–148.

Vartak V.D. (1982a). Observations on multifarious uses of sedges from western India. *J.Econ. Tax. Bot.* **3**: 193–196.

Vartak V.D.(1982b). Studies on Ethnobotany of Maharashtra and Goa. *Tribal Res.Bull.* **4(2)**: 9–10, 23.

Vartak V.D. and Madhav Gadgil (1980). Studies in ethnobotany: a new vistas in botanical sciences. *Biovigyanam* **6**: 151–156.

Vartak V.D. and Madhav Gadgil (1981). Studies on sacred goves along the western ghats from Maharashtra and Goa.: Role of Beliefs and Folklores: 272–278. In: Glimpses of Indian Ethnobotany. (Ed.S.K.Jain), Oxford and IBH Publishing Co., New Delhi.

Vartak V.D. and M.Gadgil (1990). Studies on Sacred Groves along Western Ghats in Maharashtra and Goa: Role of Belief and folklores. In: Contributions to Ethnobotany of India. 329–335.

Vartak V.D. and Rekha Mandavgane (1981). Enumeration of medicinal plants from Karnala tribal area Kolaba District, Maharashtra State. *J.Univ.Poona. (Sci.&Tech.)* **54**: 91–99.

Vartak V.D. and Vinaya Ghate (1990). Ethnobotany of Neem. *Boil.Ind.* **1(1)**: 55–59.

Vartak V.D. and V.Ghate (1994). Alu: *Meyna laxiflora* Robyns, a less known but promising wild fruits tree for tribal areas of Western Maharashtra. In higher plants of Indian Subcontinent. 3: 99–103. (Additional series of Indian Journal of Forestry No.4).

Vartak V.D.and D.K.Kulkarni (1987). Monsoon wild leafy vegetables from hilly regions of Pune and neighbouring Districts, Maharashtra State. *J.Econ.Tax.Bot.* **11(2)**: 231–235.

Vartak V.D., Kumbhojkar M.S. and D.S. Nipunge (1987). Sacred groves of tribal areas along the western Ghats: Treasure grove of medicinal plants. *B.M.E.B.R.* **8(1–2)**: 77–84.

Vartak V.D., Kumbhojkar M.S. and Vinaya S.Ghate (1985). Studies on living hedges from western Maharashtra and Goa. *J.Econ.Tax.Bot.* **6(2)**: 272–282.

Ved Prakash and B.N. Mehrotra (1987). Ethnobotanical Studies on the flora of Kandala Maharashtra State. *J.Econ. Tax.Bot.* **9(1)**: 205–208.

Wangikar B.P. (2002). Study of four edible spacies of Marchella from Aurangabad District. In: Proc. Plant Resource Development Aurangabad. 74–76. [Ed.A.M.Mungikar and A.S.Bhuktar) Dr.B.A.M. University, Aurangabad.

Watt George (1889–1893). A Dictionary of Economic Products of India Vol. 1–6. Periodical Expert, Shahadara, Delhi, India.

Yadav S.S. and P.B.Bhamare (1989). Ethnomedico–botanical studies of Dhule forests in Maharashtra. *J.Econ.Tax. Bot.* **13(2)**:455-460.

---

* Original not seen.

Appendix I

Edible Fruits & Vegetables

Edible Fruits

Botanical Name	Part Used
Aegle marmelos Correa ex Roxb.	Fruits
Anacardium occidentale L.	Fruits
Artocarpus heterophyllus Lam.	Seed
Bridellia airy-shawii P.T. Li	Fruits
Carissa congesta L.	Fruits
Cassia fistula L.	Fruits
Coccinia indica Wight & Arn.	Fruits
Cordia dichotoma Forst *f.*	Fruits
Cordia gharaf (Forssk.) Ehrenb. & Asch.	Fruits
Cucumis melo L. var.*agrestis* Naud.	Fruits
Cucumis setosus Cogn.	Fruits

Contd...

Edible Fruits–Contd...

Botanical Name	Part Used
Dendropthoe falcata (L.*f.*) Ettingshausen	Fruits
Diospyros melanoxylon Roxb.	Fruits
Ehretia laevis Roxb.	Fruits
Emblica officinalis Gaertn.	Fruits
Ficus arnottiana (Miq.) Miq.	Fruits
Ficus religiosa Linn.	Fruits
Flacourtia indica (Burm.*f.*) Merr.	Fruits
Grewia tenax (Forssk.) Foiri	Fruits
Grewia tiliaefolia Vahl	Fruits
Jasminum malabaricum Wight	Fruits
Lagascea mollis Cav.	Fruits
Lavandula bipinnata (L.) O. Klze.	Seeds
Leea asiatica L.	Fruits
Mangifera indica Linn.	Fruits
Meyna laxiflora Roybns	Fruits
Miliusa tomentosa (Roxb.) J.Sinclair	Fruits
Morinda pubescens J.E.Sm.	Fruits
Mukia maderaspatana (L.) Roem.	Fruits
Nicandra physalodes (L.) Gaertn.	Fruits
Physalis minima L.	Fruits
Phoenix sylvestris (L.) Roxb.	Fruits
Pithecelobium dulce (Roxb.) Bth.	Seed arils
Rhus sinuata Thunb.	Fruits
Securinega virosa (Roxb. ex Willd.) Pax & Hoffm.	Leaves
Sterculia urens Roxb.	Seeds
Tamarindus indica L.	Fruits
Trewia polycarpa Benth. ex Hook.*f.*	Fruits
Ziziphus nummularia (Burm. *f.*) Wight & Arn.	Fruits
Ziziphus rugosa Lam.	Fruits

Vegetables

Botanical Name	*Part Used*
Abelmoschus esculentus (Linn.) Moench.	Fruits
Alternanthera sessilis DC.	Leaves
Amaranthus spinosus L.	Leaves
Amaranthus viridis L.	Leaves
Amorphophallus commutatus (Schott.) Engl.	Tubers
Argemone mexicana L.	Seed oil
Arisaema murrayii (Grah.) Hook.	Inflorescence
Asparagus racemosus Willd.	Roots
Bombyx micranthes (L.*f.*) Riedle	Seeds
Boswellia serrata Roxb. ex Colebr.	Gum
Capparis sepiaria L.	Fruits
Capparis zeylanica L.	Fruits
Capsicum annuum Linn.	Fruits
Caralluma adscendens (Roxb.) R.Br.	Entire plant
Celosia argentea L.	Leaves
Chlorophytum tuberosum (Roxb.) Baker	Tubers
Clerodendrum serratum (L.) Moon	Leaves/Flowers
Colocasia esculenta (Linn.) Schott.	Leaves
Commelina benghalensis L.	Leaves
Commelina forskalaei Vahl	Leaves
Corchorus olitorius Linn.	Leaves
Crotolaria retusa Linn.	Flowers
Dendrocalamus strictus (Roxb.) Nees	Shoots/Buds
Digera muricata (L.) Mart.	Leaves
Dioscorea bulbifera L.	Tubers
Dioscorea hispida Dennst.	Tubers
Dioscorea oppositifolia L.	Leaves/Bulbils
Dioscorea pentaphylla Linn.	Leaves

Contd...

Vegetables–Contd...

Botanical Name	Part Used
Dioscorea wallichii Hook. *f.*	Inflorescence/Fruits
Dolichos lablab L.	Fruits
Eleusine coracana (L.) Gaertn.	Fruits
Ensete superbum (Roxb.) Cheesm.	Inflorescence
Ficus virens (Miq.) Miq.	Receptacle
Guizotia abyssinica Cass.	Seed oil
Hemigraphis latebrosa (Heyne ex Roth) Nees	Leaves
Heracaleum grandis (Dalz.) S.M.Almeida.	Fruits
Hibiscus cannabinus Linn.	Leaves
Ipomoea turbinata Lag.	Leaves
Jasminum malabaricum Wight	Fruits
Kirgenalia reticulata (Poir) Baill.	Seed oil
Indigofera cordifolia Heyne ex Roth	Seed
Ipomoea hederifolia L.	Leaves
Lagenaria siceracia Standl.	Fruits
Leucas cephalotes (Roxb. ex Roth) Spr.	Leaves
Luffa acutangula (Linn.) Roxb.	Fruits
Luffa cylindrica (L.) M.J.Roem.	Fruits
Madhuca longifolia (Koen.) MacBride var. *latifolia* (Roxb.) Chevalier	Seed oil
Marsilea minuta L.	Leaves
Momordica dioica Roxb. ex Willd.	Fruits
Mucuna prurita Hook.	Leaves
Moringa oleifera Lam.	Fruits
Musa paradisiaca Linn.	Unripe fruits
Oroxylum indicum (L.) Vent.	Seeds
Oryza sativa Linn.	Grains
Panicum miliaceum Linn.	Grains
Piliostigma malabaricum (Roxb.) Benth.	Leaves
Pimpinella heyeana Wall. ex Kurz.	Seeds
Portulaca quadrifida Linn.	Entire plant

Contd...

Vegetables–Contd...

Botanical Name	Part Used
Portulaca oleracea L.	Entire plant
Pueraria tuberosa (Willd.) DC.	Tubers
Radermachera xylocarpa (Roxb.) K.Schum.	Fruits
Rivea hypocrateriformis (Desv.) Choisy.	Leaves
Schrebrera swietenoides Roxb.	Leaves
Securinega leucopyrus (Wild.) Muell.-Arg.	Leaves
Smithia conferta J.E.Sm.	Leaves
Solanum anguivi Lam.	Leaves
Tacca leonotopetaloides (L.) O. Ktze.	Tubers
Tricholepis glaberrima DC.	Leaves
Triumfetta rhomboidea Jacq.	Leaves
Urena lobata L.	Leaves
Vigna radiata (L.) Wilczek, v.*sulobata* (Roxb.) Verdc.	Seeds

Appendix II (A)

Plants Employed for House Construction

Acacia nilotica (Linn.) Delile Subsp. indica (Benth.) Brenan

Agave vera-cruz Mill.

Albizzia lebbeck Benth.

Artocarpus heterophyllus Lam.

Bambusa arundinacea (Retz.) Willd.

Bambusa vulgaris Schrad.

Bauhinia racemosa Lam.

Borassus flabelliformis Linn.

Erythrina stricta Roxb.

Ficus benghalensis Linn.

Gmelina arborea Roxb.

Gliricidia sepium (Jacq.) Kunth ex Steud.

Holarrhena pubescens (Linn.) Wall.

Lagerstroemia parviflora Roxb.

Lannea coromandelica (Houtt.) Merrill

Mitragyna parviflora (Roxb.) Kirth.

Morinda pubescens J.E.Sm.

Phoenix sylvestris (L.) Roxb.

Securinega leucopyrus (Willd.) Muell.-Arg.

Tectona grandis L.

Trichosanthes tricupsidata Lour.

Appendix II (B)

Plants Employed for Hut Thatching

Apluda mutica Linn.

Aegle marmelos Correa ex Roxb.

Cajanus Cajan (Linn.) Millsp.

Carvia callosa (Nees) Bremek.

Colebrookia oppositifolia J.E.Sm.

Cyperus pangorei Rottb.

Cyperus rotundus L.

Dichanthium pertusum (L.) Clayton

Eriophorum comosum Wall.

Ipomoea carnea Jacq.

Lantana camara Linn.

Madhuca longifolia (Koen) MacBride var. *latifolia* (Roxb.) Chevalier

Schleichera oleosa (Lour.) Oken

Sorghum halepense (L.) Pers.

Sorghum bicolor (L.) Moench.

Vitex negundo L.

Appendix III
Plants Used for Hedges

Agave vera-cruz Mill.

Caesalpinia decapetala (Roth) Alst.

Carvia callosa (Nees) Bremek.

Dendrocalamus strictus (Roxb.) Nees.

Euphorbia neirifolia Linn.

Lantana camara Linn.

Maytenus senegalensis (Lamk.) Exell.

Prosopis juliflora DC.

Securinega leucopyrus (Willd.) Muell.-Arg

Vitex negundo L.

Appendix IV

Plants Used as Fodder

TREES

Acacia leucophloea Willd.

Acacia nilotica (Linn.) Delile subsp. indica (Benth.) Brenan

Albizia lebbeck Benth.

Bauhinia racemosa Lam.

Careya arborea Roxb.

Cordia dichotoma Forst. f.

Ficus amplissima J.E.Sm.

Ficus arnottiana (Miq.) Miq.

Ficus racemosa L.

Ficus religiosa Linn.

Ficus virens (Miq.) Miq.

Flacourtia indica (Burm. *f.*) Merr.

Gmelina arborea Roxb.

Hardwickia binata Roxb.

Ixora pavetta Andr.

Mimosa hamata Willd.

Mitragyna parvifolia (Roxb.) Korth.

Ougeinia oojeinensis (Roxb.) Hochr.

Pithecelobium dulce (Roxb.) Bth.

Prosopis juliflora DC.

Terminalia chebula Retz.

Ziziphus mauritiana Lam.

Ziziphus nummularia (Burm.*f.*) Wight & Arn.

Ziziphus rugosa Lam.

HERBS, SHRUBS AND CLIMBERS

**Acalypha malabarica* Muell.-Arg.

Aerva lanata (L.) A.L.Juss. ex Schult.

Aeschynomene indica L.

Alternanthera sessilis DC.

**Alysicarpus rugosus* (Willd.) DC.

Amaranthus spinosus L.

**Amaranthus viridis* L.

Apluda mutica Linn.

Arthaxon lancifolius (Trin.) Hoschst.

Aspidopterys cordata (Heyne ex Wall.) A. Juss.

Bambusa arundinacea (Retz.) Willd.

**Barleria lawii* T. Andr.

Bauhinia racemosa Lam.

Caesalpinia decapetala (Roth) Alst.

Capparis spinosa L.

Cardiospermum halicacabum L.

Cassia tora Linn.

**Celosia argentea* L.

Chlorophytum tuberosum (Roxb.) Baker

Clerodendrum multiflorum (Burm. *f.*) O.Ketz.

Clitoria ternatea Linn.

**Cocculus hirsutus* DC.

Commelina benghalensis L.

**Commelina forskalaei* Vahl

**Commelina hasskarlii* C.B.Cl.

Commicarpus plumbagineus (Cav.) Standl.

Commicarpus chinensis (L.) Heimerl.

**Corchorus olitorius* Linn.

**Corchorus trilocularis* Linn.

**Crotalaria filipes* Bth.

Ctenolepis garcinii (Burm. *f.*) C.B.Cl.

Cucumis setosus Cogn.

Cymbopogon martinii (Roxb.) Wats.

**Cynodon dactylon* (L.) Pers.

**Desmodium dichotomum* (Willd.) DC.

Dichrostachys cinerea (L.) Wight & Arn.

Dinebra retroflexa (Vahl) Panz.

**Digera muricata* (L.) Mart.

Dolichus lablab L.

Euphorbia geniculata Orteg.

**Euphorbia parviflora* L.

Eranthemum roseum Linn.

Grewia tiliaefolia Vahl

Guizotia abyssinica (L.*f.*) Cass.

Hemigraphis latebrosa (Heyne ex Roth) Nees

Hibiscus cannabinus Linn.

Ipomoea hederifolia L.

Ipomoea turbinata Lag.

**Launea procumbens* (Roxb.) Ramayya and Rajgopal

**Leucas cephalotes* (Roxb. ex Roth) Spr.

Maytenus senegalensis (Lamk.) Excell.

Pergularia daemia (Forsk.) Chiov.

**Phyllanthus fraternus* Webster

**Physalis minima* L.

**Portulaca quadrifida* Linn.

Prosopis cineraria (L.) Druce

Psoralea corylifolia L.

Pupalia lappacea (L.) A.L.Juss.

Rhus sinuata Thunb.

Rhyncosia minima DC.

Rivea hypocrateriformis (Desv.) Choisy.

**Setaria pumila*

Sida rhombifolia L.

**Sonchus oleraceus L.*

Sorghum bicolor (L.) Moench.

Vigna radiata (L.) Wilczek.

Wattakaka volubilis (L.*f.*) Stap *f.*

Ziziphus nummularia (Burm. *f.*) Wight & Arn.

---

*Indicates crop weeds

Appendix V

Plants Used for Fish Surveying

Botanical Name	Part Used
Acacia auriculiformis A.Cunn.	Fruits
Catunaregum spinosa (Thunb.) Tiruvengadum	Fruits
Costus speciosus (Koen.) J.E.Sm.	Tubers
Erythrina suberosa Roxb.	Bark
Garuga pinnata Roxb.	Fruits
Holoptelea integrifolia (Roxb.) Planch	Leaves
Hollarrhena pubescens (Linn.) Wall.	Twigs
Madhuca longifolia (Koen.) MacBride var.*latifolia* (Roxb.) Chevalier	Stem
Plectranthes mollis Spreng.	Leaves
Sterculia urens Roxb.	Leaves
Terminalia bellriica Roxb.	Fruits
Wrightia tinctoria R.Br.	Bark

Appendix VI
Plants Used for Veterinary Medicine

Botanical Name	*Part/Product Used*
Acacia auriculiformis A.Cunn.	Fruits
Adansonia digitata L.	Fruits
Agave cantala Roxb.	Leaves
Ailanthes excelsa Roxb.	Leaves
Annona squamosa L.	Leaves
Argyeria sericea Dalz.	Leaves
Asparagus racemosus Willd.	Tubers
Aspidopterys cordata (Heyne ex Wall.) A.Juss.	Leaves
Balanites aegyptiaca (L.) Delile	Stem
Bambusa vulgaris Schrad.	Stem
Bauhinia vahlii W.&A.	Bark

Contd...

Botanical Name	Part/Product Used
Butea monosperma (Lam.) Taub.	Twigs
Butea superba Roxb.	Roots
Calotropis gigantea (Linn.) R.Br.ex Ait.	Latex
Calotropis procera (Ait.) R.Br.	Leaf buds
Carissa congesta L.	Roots
Carum carvi Linn.	Fruits
Cassia fistula L.	Stem
Cassine albens *(Retz.) Koesterm.*	Stem/Bark
Cassia auriculata (Wall. ex Wight & Arn.) Gamble	Stem
Cayratia trifolia (L.) Domin	Twigs
Celosia argentea L.	Root
Cissus quadrangularis L.	Stem
Cissus rependa Vahl	Stem
Clematis triloba Hyeyne ex Roth	Twigs
Canscora diffusa (Vahl) R.Br.	Plant Ash
Curcuma domestica Valeton	Rhizome
Cuscuta reflexa Roxb.	Stem
Dendrocalamus strictus (Roxb.) Nees	Leaves
Desmodium gangetcum DC.	Leaves
Dioscorea hispida Dennst.	Tubers
Diospyros melanoxylon Roxb.	Stem
Diplocyclos palmatus (L.) Jeffery	Plant
Dolichandrone falcata Seem.	Bark/Poultry
Echinops echinatus Roxb.	Root
Ensete superbum (Roxb.) Cheesm.	Seeds/Poultry
Erythrina variegata Linn.	Bark
Eranthemum roseum Linn.	Roots
Euphorbia fusiformis Buch.-Ham. ex D.Don	Latex
Ficus arnottiana (Miq.) Miq.	Latex
Gloriosa superba L.	Tubers
Grewia tenax (Forssk.) Fiori.	Root
Hamiltonia suaveolens Roxb.	Bark

Contd...

Botanical Name	*Part/Product Used*
Hardwickia binnata Roxb.	Leaves
Heracleum grandis (Dalz.) S.M.Almeida	Roots
Holarrhena pubescens (Linn.) Wall.	Bark
Holoptelea integrifolia (Roxb.) Planch	Fibres
Jatropha curcas L.	Twigs
Lagestroemia parviflora Roxb.	Leaves
Leea macrophylla Roxb. ex Horn.	Roots
Madhuca longifolia (Koen.) MacBride var. *latifolia* (Roxb.) Chevalier	Seeds
Nervilia plicata (Andr.) Schlecht.	Bulb
Oroxylum indicum Vent.	Bark
Pistia stratiotes Linn.	Leaves
Pterocarpus marsupium Roxb.	Leaves
Pureria tuberosa (Willd.) DC.	Tubers
Tamarindus indica Linn.	Leaves
Tamilnadia uliginosa (Retz.) Trivengadum & Sastre	Bark
Terminalia bellirica (Gaertn.) Roxb.	Fruits/Poultry
Wrightia tinctoria R.Br.	Twigs

Appendix VII

Antidotes

Botanical Name	Part Used
SCORPION STING	
Achyranthes aspera L.	Roots
Anacardium occidentale L.	Seeds
Anogeissus latifolia (Roxb. ex DC.) Wall. ex Guill. & Perr.	Bark
Dalbergia lanceolaria L.*f.*	Bark
Echinops echinatus Roxb.	Roots
Tridax procumbens Linn.	Leaves
SNAKE BITE	
Acacia polycantha Willd.	Bark
Azadirachta indica A.Juss.	Leaves
Casearia esculenta Roxb.	Stem
Celosia argentea L.	Root
Cissampelos pariera Linn.	Leaves/Bark
Coix gigantea Koen. ex Roxb.	Root
Cucumis prophetarum Linn.	Fruits

Contd...

Botanical Name	Part Used
Emblica officinalis Gaertn.	Bark
Holarrhena pubescens (Linn.) Wall.	Bark
Kedrostris rostrata (Rott.) Cogn.	Roots
Leea macrophylla Roxb. ex Horn.	Roots
Melia composita Willd.	Bark
Momordica dioica Roxb. ex Willd.	Root tubers
Phaseolus lunatus Linn.	Seed
Pogostemon parviflora Bth.	Root
Trichosanthes tricupsidata Lour. (Willd.) Miers ex Hook.*f.* & Thoms.	Stem/Fruits
Tinospora cordifolia (Willd.) Miers ex Hook. *f.* & Thoms.	Stem/leaves
Vigna trilobata (Linn.) Verdcourt	Tuber
Wrightia tinctoria R.Br.	Bark
Woodfordia fruticosa (L.) Kurz.	Leaves

Appendix VIII
Plants in Ethnobeliefs

Botanical Name	*Part Used*
Adansonia digitata L.	Fruit
Adiantum philipense Linn.	Stem
Bauhinia tomentosa Linn.	Stem
Butea monosperma (Lam.) Taub.	Twigs
Casearia esculenta Roxb.	Stem
Cassia fistula L.	Pods/Stem
Cassine albens (Retz.) Koesterm.	Stem
Cissus quadrangularis Linn.	Stem
Cissus rependa Vahl	Climber
Cryptolepis buchanani R. & S.	Twig
Dendropthoe falcata (L.*f.*) Ettingshausen	Stem
Desmodium gangeticum (L.) DC.	Leaves

Contd...

Botanical Name	Part Used
Diplocyclos palmatus (L.) Jeffery	Seed
Holarrhena pubescens (Linn.) Wall.	Twig
Kyllinga tenuifolia Steud.	Plant
Lepiosorus thunbergianus (Kaulf.) Ching	Stem
Mucuna prurita Hook.	Fruits
Pandanus odoratissimus Linn. *f.*	Bracts
Santalum album L.	Wood
Urena lobata L.	Stem

Appendix IX

Plants Used for Religious Ceremonies

Botanical Name	Part Used
Aegle marmelos Correa ex Roxb.	Leaves
Calotropis gigantea (Linn.) R.Br. ex Ait.	Leaves/Flowers
Curcuma pseudomontana Grah.	Inflorescence
Cynodon dactylon (L.) Pers.	Plant
Cyperus pangorei Rottb.	Stem
Gloriosa superba L.	Flowers
Guizotia abyssinica (L.*f.*) Cass.	Flowers
Mangifera indica Linn.	Leaves
Musa paradisiaca Linn.	Inflorescence
Tectona grandis L.	Wood
Vicoa indica (L.) DC.	Stem

Appendix X

Plants Used for Agricultural Implements

Botanical Name	Part Used
Acacia ferruginea DC.	Wood
Acacia leucophloea Willd.	Wood
Acacia nilotica (Linn.) Delile subsp. indica (Benth.) Brenan.	Wood
Cordia dichotoma Forst. *f.*	Wood
Dichrostachys cinerea (L.) Wight & Arn.	Wood
Hardwickia binata Roxb.	Wood
Heterophragma quadriloculare (Roxb.) K. Schum.	Wood
Mangifera indica Linn.	Wood
Memecylon umbellatum Burm.*f.*	Wood
Mimosa hamata Willd.	Wood

Contd...

Botanical Name	Part Used
Ougeinia oojeinensis (Roxb.) Hochr.	Wood
Tectona grandis L.f.	Wood
Terminalia crenulata Roth	Wood
Terminalia chebula Retz.	Wood
Typha australis Schum. & Thonn.	Leaves
Vitex negundo L.	Wood

Appendix XI

Plants Used for Fire Wood

Albizia lebbeck (L.) Bth.

Ampelocissus latifolia (Roxb.) Planch.

Anogeissus latifolia (Roxb. ex DC.) Wall. ex Guill & Perr.

Caesalpinia sepiaria Roxb.

Carissa congesta Wight

Colebrookea oppositifolia J.E.Sm.

Flacourtia indica (Burm. *f.*) Merr.

Gliricidia sepium (Jacq.) Kunth ex Steud.

Gmelina arborea Roxb.

Lantana camara L.

Miliusa tomentosa (Roxb.) Sinclair

Phoenix sylvestris (L.) Roxb.

Syzigium cumunii (L.) Skeels

Tectona grandis L.*f.*

Terminalia bellirica (Gaertn.) Roxb.

Terminalia crenulata Roth

Vitex negundo L.

Appendix XII
Crop Weeds

Acalypha malabarica Muell-Arg.

**Ageratum conyzoides* Linn.

Alysicarpus rugosus (Willd.) DC.

Amaranthus spinosus L.

**Argemone mexicana* L.

Brachiara reptans (Linn.) Gard & C.E. Hubbard

Celosia argentea L.

Cocculus hirsutus DC.

**Corchorus aestuans* Linn.

Corchorus olitorius Linn.

Corchorus trilocularis Linn.

Cynodon dactylon (L.) Pers.

Cyperus rotundus L.

Desmodium dichotomum (Willd.) DC.

**Digera muricata* (L.) Mart.

Echinops echinatus Roxb.

Emilia sonchifolia (L.) DC.

Euphorbia parviflora L.

Launea procumbens (Roxb.) Ramayya and Rajagopal

Leucas cephalotes (Roxb. ex Roth) Spr.

**Oxalis corniculata* Linn.

Phyllanthus fraternus Webster.

**Physalis minima* L.

Portulaca quadrifida Linn.

**Portulaca oleracea* L.

Rhyncosia minima DC.

Rungia repens (L.) Nees

Setaria pumila (Poir.) R. & S.

Solanum surattense Burm. f.

Solena virginianum L.

Trichodesma indicum (L.) Br.

Tricholepis glaberrima DC.

**Tridax procumbens* L.

Vicoa indica DC.

---

* Indicate exotic crop-weeds

Appendix XIII

Exotic Plants

Botanical Name	*Native Country*
Abelmoschus esculentus (Linn.) Moench.	Tropical Africa
Acacia auriculiformis A.Cunn. ex Benth.	Australia
Acacia nilotica (L.) Willd.	North Africa
Achyranthes aspera L.	North Africa
Adansonia digitata L.	Tropical Africa
Agave cantala Roxb.	America
Agave vera-cruz Mill.	American Desert
Ageratum conyzoides L.	South America
Alternanthera tenella Colla	Tropical America
Aloe vera (L.) Burm. *f.*	North America
Anacardium occidentale L.	Tropical America
Annona squamosa L.	Tropical America
Argemone mexicana L.	Mexico
Azadirachta indica A.Juss.	Persia
Cajanus cajan (Linn.) Millsp.	Africa
Capsicum annum L.	Tropical America

Contd...

Botanical Name	Native Country
Cassia occidentalis L.	South America
Cassia tora L.	America
Cissampelos pareira L.	South America
Clitoria ternatea L.	Tropical America
Corchorus aestuans L.	Tropical America
Digera muricata (L.) Mart.	North Africa
Gliricidia sepium (Jacq.) Kunth ex Steud.	Central America
Guizotia abyssinica (L.*f.*) Cass.	Abyssinia
Helianthus annuus Linn.	North America
Hibiscus cannabinus L.	Tropical America
Ipomoea carnea Jacq.	South America
Jatropha curcas L.	Tropical America
Lantana camara Linn.	Tropical America
Nicandra physalodes (L.) Gaertn.	Peru
Opuntia dillenii (Ker.-Gawl.) Haw.	South America
Physalis minima L.	South America
Oxalis corniculata Linn.	South Europe/North America
Pithecelobium dulce (Roxb.) Bth.	Mexico
Plumbago zeylanica L.	Africa
Portulaca oleracea L.	North Africa
Prosopis juliflora (Sw.) DC.	Tropical America
Ricinus communis L.	Africa
Sonchus oleraceus L.	Europe
Sorghum halepense (L.) Pers.	Arabia
Sorghum bicolor (L.) Moench.	Tropical America
Tagetes erecta Linn.	Africa/Mexico
Tamarindus indica L.	Tropical Africa
Tridax procumbens L.	South America
Urena lobata L.	Probably Africa
Xanthium strumarium L.	Europe

Appendix XIV

Plants Reported Useful for First Time

Aerides crispum Lindl.

Arthraxon lancifolius (Trin.) Hoschst.

Argyeria sericea Dalz.

Barleria lawii T. Andr.

Alternanthera tenella Colla

Commelina forskalaei Vahl

Commelina hasskarlii C.B.Cl.

Commicarpus plumbagineus (Cav.) Standl.

Crotalaria filipes Bth.

Ctenolepis garcinii (Burm.*f.*) C.B.Cl.

Cucumis setosus Cogn.

Dendrobium barbatulum Lindl.

Euphorbia geniculata Orteg.

Gliricidia sepium (Jacq.) Walp.

Haplanthodes verticillatus (Roxb.) Nees

Hemigraphis latebrosa (Heyne ex Roth) Nees

Ipomoea hederifolia L.

Kyllinga tenuifolia Steud.

Nervilia plicata (Andr.) Schlecht.

Pancratium sanctae-mariae Blatt. & Haub.

Setaria pumila (Poir.) R.&S.

Index to Local Names

Index to Plant Names

A